Heat Treatment for Insect Control

Related titles

Hygiene in food processing
(ISBN 978-0-85709-429-2)

Hygienic design of food factories
(ISBN 978-1-84569-564-4)

Handbook of hygiene control in the food industry
(ISBN 978-1-85573-957-4)

Woodhead Publishing Series in Food Science, Technology and Nutrition: Number 241

Heat Treatment for Insect Control

Developments and Applications

David Hammond

ELSEVIER

AMSTERDAM • BOSTON • CAMBRIDGE • HEIDELBERG
LONDON • NEW YORK • OXFORD • PARIS • SAN DIEGO
SAN FRANCISCO • SINGAPORE • SYDNEY • TOKYO
Woodhead Publishing is an imprint of Elsevier

WP WOODHEAD PUBLISHING

Woodhead Publishing is an imprint of Elsevier
80 High Street, Sawston, Cambridge, CB22 3HJ, UK
225 Wyman Street, Waltham, MA 02451, USA
Langford Lane, Kidlington, OX5 1GB, UK

British Library Cataloguing-in-Publication Data
A catalogue record for this book is available from the British Library

Library of Congress Control Number: 2014941275

ISBN 978-0-85709-776-7 (print)
ISBN 978-0-85709-781-1 (online)

For information on all Woodhead Publishing publications
visit our website at http://store.elsevier.com/

Typeset by RefineCatch Limited, Bungay, Suffolk

Contents

Woodhead Publishing Series in Food Science, Technology and Nutrition

1 **Chilled foods: A comprehensive guide**
Edited by C. Dennis and M. Stringer
2 **Yoghurt: Science and technology**
A. Y. Tamime and R. K. Robinson
3 **Food processing technology: Principles and practice**
P. J. Fellows
4 **Bender's dictionary of nutrition and food technology Sixth edition**
D. A. Bender
5 **Determination of veterinary residues in food**
Edited by N. T. Crosby
6 **Food contaminants: Sources and surveillance**
Edited by C. Creaser and R. Purchase
7 **Nitrates and nitrites in food and water**
Edited by M. J. Hill
8 **Pesticide chemistry and bioscience: The food-environment challenge**
Edited by G. T. Brooks and T. Roberts
9 **Pesticides: Developments, impacts and controls**
Edited by G. A. Best and A. D. Ruthven
10 **Dietary fibre: Chemical and biological aspects**
Edited by D. A. T. Southgate, K. W. Waldron, I. T. Johnson and G. R. Fenwick
11 **Vitamins and minerals in health and nutrition**
M. Tolonen
12 **Technology of biscuits, crackers and cookies Second edition**
D. Manley
13 **Instrumentation and sensors for the food industry**
Edited by E. Kress-Rogers
14 **Food and cancer prevention: Chemical and biological aspects**
Edited by K. W. Waldron, I. T. Johnson and G. R. Fenwick
15 **Food colloids: Proteins, lipids and polysaccharides**
Edited by E. Dickinson and B. Bergenstahl
16 **Food emulsions and foams**
Edited by E. Dickinson
17 **Maillard reactions in chemistry, food and health**
Edited by T. P. Labuza, V. Monnier, J. Baynes and J. O'Brien
18 **The Maillard reaction in foods and medicine**
Edited by J. O'Brien, H. E. Nursten, M. J. Crabbe and J. M. Ames
19 **Encapsulation and controlled release**
Edited by D. R. Karsa and R. A. Stephenson
20 **Flavours and fragrances**
Edited by A. D. Swift
21 **Feta and related cheeses**
Edited by A. Y. Tamime and R. K. Robinson
22 **Biochemistry of milk products**
Edited by A. T. Andrews and J. R. Varley
23 **Physical properties of foods and food processing systems**
M. J. Lewis
24 **Food irradiation: A reference guide**
V. M. Wilkinson and G. Gould

25 **Kent's technology of cereals: An introduction for students of food science and agriculture Fourth edition**
N. L. Kent and A. D. Evers
26 **Biosensors for food analysis**
Edited by A. O. Scott
27 **Separation processes in the food and biotechnology industries: Principles and applications**
Edited by A. S. Grandison and M. J. Lewis
28 **Handbook of indices of food quality and authenticity**
R. S. Singhal, P. K. Kulkarni and D. V. Rege
29 **Principles and practices for the safe processing of foods**
D. A. Shapton and N. F. Shapton
30 **Biscuit, cookie and cracker manufacturing manuals Volume 1: Ingredients**
D. Manley
31 **Biscuit, cookie and cracker manufacturing manuals Volume 2: Biscuit doughs**
D. Manley
32 **Biscuit, cookie and cracker manufacturing manuals Volume 3: Biscuit dough piece forming**
D. Manley
33 **Biscuit, cookie and cracker manufacturing manuals Volume 4: Baking and cooling of biscuits**
D. Manley
34 **Biscuit, cookie and cracker manufacturing manuals Volume 5: Secondary processing in biscuit manufacturing**
D. Manley
35 **Biscuit, cookie and cracker manufacturing manuals Volume 6: Biscuit packaging and storage**
D. Manley
36 **Practical dehydration Second edition**
M. Greensmith
37 **Lawrie's meat science Sixth edition**
R. A. Lawrie
38 **Yoghurt: Science and technology Second edition**
A. Y. Tamime and R. K. Robinson
39 **New ingredients in food processing: Biochemistry and agriculture**
G. Linden and D. Lorient
40 **Benders' dictionary of nutrition and food technology Seventh edition**
D. A. Bender and A. E. Bender
41 **Technology of biscuits, crackers and cookies Third edition**
D. Manley
42 **Food processing technology: Principles and practice Second edition**
P. J. Fellows
43 **Managing frozen foods**
Edited by C. J. Kennedy
44 **Handbook of hydrocolloids**
Edited by G. O. Phillips and P. A. Williams
45 **Food labelling**
Edited by J. R. Blanchfield
46 **Cereal biotechnology**
Edited by P. C. Morris and J. H. Bryce
47 **Food intolerance and the food industry**
Edited by T. Dean
48 **The stability and shelf-life of food**
Edited by D. Kilcast and P. Subramaniam
49 **Functional foods: Concept to product**
Edited by G. R. Gibson and C. M. Williams
50 **Chilled foods: A comprehensive guide Second edition**
Edited by M. Stringer and C. Dennis
51 **HACCP in the meat industry**
Edited by M. Brown
52 **Biscuit, cracker and cookie recipes for the food industry**
D. Manley

53 **Cereals processing technology**
Edited by G. Owens
54 **Baking problems solved**
S. P. Cauvain and L. S. Young
55 **Thermal technologies in food processing**
Edited by P. Richardson
56 **Frying: Improving quality**
Edited by J. B. Rossell
57 **Food chemical safety Volume 1: Contaminants**
Edited by D. Watson
58 **Making the most of HACCP: Learning from others' experience**
Edited by T. Mayes and S. Mortimore
59 **Food process modelling**
Edited by L. M. M. Tijskens, M. L. A. T. M. Hertog and B. M. Nicolaï
60 **EU food law: A practical guide**
Edited by K. Goodburn
61 **Extrusion cooking: Technologies and applications**
Edited by R. Guy
62 **Auditing in the food industry: From safety and quality to environmental and other audits**
Edited by M. Dillon and C. Griffith
63 **Handbook of herbs and spices Volume 1**
Edited by K. V. Peter
64 **Food product development: Maximising success**
M. Earle, R. Earle and A. Anderson
65 **Instrumentation and sensors for the food industry Second edition**
Edited by E. Kress-Rogers and C. J. B. Brimelow
66 **Food chemical safety Volume 2: Additives**
Edited by D. Watson
67 **Fruit and vegetable biotechnology**
Edited by V. Valpuesta
68 **Foodborne pathogens: Hazards, risk analysis and control**
Edited by C. de W. Blackburn and P. J. McClure
69 **Meat refrigeration**
S. J. James and C. James
70 **Lockhart and Wiseman's crop husbandry Eighth edition**
H. J. S. Finch, A. M. Samuel and G. P. F. Lane
71 **Safety and quality issues in fish processing**
Edited by H. A. Bremner
72 **Minimal processing technologies in the food industries**
Edited by T. Ohlsson and N. Bengtsson
73 **Fruit and vegetable processing: Improving quality**
Edited by W. Jongen
74 **The nutrition handbook for food processors**
Edited by C. J. K. Henry and C. Chapman
75 **Colour in food: Improving quality**
Edited by D. MacDougall
76 **Meat processing: Improving quality**
Edited by J. P. Kerry, J. F. Kerry and D. A. Ledward
77 **Microbiological risk assessment in food processing**
Edited by M. Brown and M. Stringer
78 **Performance functional foods**
Edited by D. Watson
79 **Functional dairy products Volume 1**
Edited by T. Mattila-Sandholm and M. Saarela
80 **Taints and off-flavours in foods**
Edited by B. Baigrie
81 **Yeasts in food**
Edited by T. Boekhout and V. Robert
82 **Phytochemical functional foods**
Edited by I. T. Johnson and G. Williamson

83 **Novel food packaging techniques**
Edited by R. Ahvenainen
84 **Detecting pathogens in food**
Edited by T. A. McMeekin
85 **Natural antimicrobials for the minimal processing of foods**
Edited by S. Roller
86 **Texture in food Volume 1: Semi-solid foods**
Edited by B. M. McKenna
87 **Dairy processing: Improving quality**
Edited by G. Smit
88 **Hygiene in food processing: Principles and practice**
Edited by H. L. M. Lelieveld, M. A. Mostert, B. White and J. Holah
89 **Rapid and on-line instrumentation for food quality assurance**
Edited by I. Tothill
90 **Sausage manufacture: Principles and practice**
E. Essien
91 **Environmentally-friendly food processing**
Edited by B. Mattsson and U. Sonesson
92 **Bread making: Improving quality**
Edited by S. P. Cauvain
93 **Food preservation techniques**
Edited by P. Zeuthen and L. Bøgh-Sørensen
94 **Food authenticity and traceability**
Edited by M. Lees
95 **Analytical methods for food additives**
R. Wood, L. Foster, A. Damant and P. Key
96 **Handbook of herbs and spices Volume 2**
Edited by K. V. Peter
97 **Texture in food Volume 2: Solid foods**
Edited by D. Kilcast
98 **Proteins in food processing**
Edited by R. Yada
99 **Detecting foreign bodies in food**
Edited by M. Edwards
100 **Understanding and measuring the shelf-life of food**
Edited by R. Steele
101 **Poultry meat processing and quality**
Edited by G. Mead
102 **Functional foods, ageing and degenerative disease**
Edited by C. Remacle and B. Reusens
103 **Mycotoxins in food: Detection and control**
Edited by N. Magan and M. Olsen
104 **Improving the thermal processing of foods**
Edited by P. Richardson
105 **Pesticide, veterinary and other residues in food**
Edited by D. Watson
106 **Starch in food: Structure, functions and applications**
Edited by A-C. Eliasson
107 **Functional foods, cardiovascular disease and diabetes**
Edited by A. Arnoldi
108 **Brewing: Science and practice**
D. E. Briggs, P. A. Brookes, R. Stevens and C. A. Boulton
109 **Using cereal science and technology for the benefit of consumers: Proceedings of the 12th International ICC Cereal and Bread Congress, 24–26th May, 2004, Harrogate, UK**
Edited by S. P. Cauvain, L. S. Young and S. Salmon
110 **Improving the safety of fresh meat**
Edited by J. Sofos
111 **Understanding pathogen behaviour: Virulence, stress response and resistance**
Edited by M. Griffiths

112 **The microwave processing of foods**
Edited by H. Schubert and M. Regier
113 **Food safety control in the poultry industry**
Edited by G. Mead
114 **Improving the safety of fresh fruit and vegetables**
Edited by W. Jongen
115 **Food, diet and obesity**
Edited by D. Mela
116 **Handbook of hygiene control in the food industry**
Edited by H. L. M. Lelieveld, M. A. Mostert and J. Holah
117 **Detecting allergens in food**
Edited by S. Koppelman and S. Hefle
118 **Improving the fat content of foods**
Edited by C. Williams and J. Buttriss
119 **Improving traceability in food processing and distribution**
Edited by I. Smith and A. Furness
120 **Flavour in food**
Edited by A. Voilley and P. Etievant
121 **The Chorleywood bread process**
S. P. Cauvain and L. S. Young
122 **Food spoilage microorganisms**
Edited by C. de W. Blackburn
123 **Emerging foodborne pathogens**
Edited by Y. Motarjemi and M. Adams
124 **Benders' dictionary of nutrition and food technology Eighth edition**
D. A. Bender
125 **Optimising sweet taste in foods**
Edited by W. J. Spillane
126 **Brewing: New technologies**
Edited by C. Bamforth
127 **Handbook of herbs and spices Volume 3**
Edited by K. V. Peter
128 **Lawrie's meat science Seventh edition**
R. A. Lawrie in collaboration with D. A. Ledward
129 **Modifying lipids for use in food**
Edited by F. Gunstone
130 **Meat products handbook: Practical science and technology**
G. Feiner
131 **Food consumption and disease risk: Consumer–pathogen interactions**
Edited by M. Potter
132 **Acrylamide and other hazardous compounds in heat-treated foods**
Edited by K. Skog and J. Alexander
133 **Managing allergens in food**
Edited by C. Mills, H. Wichers and K. Hoffman-Sommergruber
134 **Microbiological analysis of red meat, poultry and eggs**
Edited by G. Mead
135 **Maximising the value of marine by-products**
Edited by F. Shahidi
136 **Chemical migration and food contact materials**
Edited by K. Barnes, R. Sinclair and D. Watson
137 **Understanding consumers of food products**
Edited by L. Frewer and H. van Trijp
138 **Reducing salt in foods: Practical strategies**
Edited by D. Kilcast and F. Angus
139 **Modelling microorganisms in food**
Edited by S. Brul, S. Van Gerwen and M. Zwietering
140 **Tamime and Robinson's Yoghurt: Science and technology Third edition**
A. Y. Tamime and R. K. Robinson
141 **Handbook of waste management and co-product recovery in food processing Volume 1**
Edited by K. W. Waldron

142 **Improving the flavour of cheese**
Edited by B. Weimer
143 **Novel food ingredients for weight control**
Edited by C. J. K. Henry
144 **Consumer-led food product development**
Edited by H. MacFie
145 **Functional dairy products Volume 2**
Edited by M. Saarela
146 **Modifying flavour in food**
Edited by A. J. Taylor and J. Hort
147 **Cheese problems solved**
Edited by P. L. H. McSweeney
148 **Handbook of organic food safety and quality**
Edited by J. Cooper, C. Leifert and U. Niggli
149 **Understanding and controlling the microstructure of complex foods**
Edited by D. J. McClements
150 **Novel enzyme technology for food applications**
Edited by R. Rastall
151 **Food preservation by pulsed electric fields: From research to application**
Edited by H. L. M. Lelieveld and S. W. H. de Haan
152 **Technology of functional cereal products**
Edited by B. R. Hamaker
153 **Case studies in food product development**
Edited by M. Earle and R. Earle
154 **Delivery and controlled release of bioactives in foods and nutraceuticals**
Edited by N. Garti
155 **Fruit and vegetable flavour: Recent advances and future prospects**
Edited by B. Brückner and S. G. Wyllie
156 **Food fortification and supplementation: Technological, safety and regulatory aspects**
Edited by P. Berry Ottaway
157 **Improving the health-promoting properties of fruit and vegetable products**
Edited by F. A. Tomás-Barberán and M. I. Gil
158 **Improving seafood products for the consumer**
Edited by T. Børresen
159 **In-pack processed foods: Improving quality**
Edited by P. Richardson
160 **Handbook of water and energy management in food processing**
Edited by J. Klemeš, R. Smith and J-K. Kim
161 **Environmentally compatible food packaging**
Edited by E. Chiellini
162 **Improving farmed fish quality and safety**
Edited by Ø. Lie
163 **Carbohydrate-active enzymes**
Edited by K-H. Park
164 **Chilled foods: A comprehensive guide Third edition**
Edited by M. Brown
165 **Food for the ageing population**
Edited by M. M. Raats, C. P. G. M. de Groot and W. A Van Staveren
166 **Improving the sensory and nutritional quality of fresh meat**
Edited by J. P. Kerry and D. A. Ledward
167 **Shellfish safety and quality**
Edited by S. E. Shumway and G. E. Rodrick
168 **Functional and speciality beverage technology**
Edited by P. Paquin
169 **Functional foods: Principles and technology**
M. Guo
170 **Endocrine-disrupting chemicals in food**
Edited by I. Shaw
171 **Meals in science and practice: Interdisciplinary research and business applications**
Edited by H. L. Meiselman

172 **Food constituents and oral health: Current status and future prospects**
Edited by M. Wilson
173 **Handbook of hydrocolloids Second edition**
Edited by G. O. Phillips and P. A. Williams
174 **Food processing technology: Principles and practice Third edition**
P. J. Fellows
175 **Science and technology of enrobed and filled chocolate, confectionery and bakery products**
Edited by G. Talbot
176 **Foodborne pathogens: Hazards, risk analysis and control Second edition**
Edited by C. de W. Blackburn and P. J. McClure
177 **Designing functional foods: Measuring and controlling food structure breakdown and absorption**
Edited by D. J. McClements and E. A. Decker
178 **New technologies in aquaculture: Improving production efficiency, quality and environmental management**
Edited by G. Burnell and G. Allan
179 **More baking problems solved**
S. P. Cauvain and L. S. Young
180 **Soft drink and fruit juice problems solved**
P. Ashurst and R. Hargitt
181 **Biofilms in the food and beverage industries**
Edited by P. M. Fratamico, B. A. Annous and N. W. Gunther
182 **Dairy-derived ingredients: Food and neutraceutical uses**
Edited by M. Corredig
183 **Handbook of waste management and co-product recovery in food processing Volume 2**
Edited by K. W. Waldron
184 **Innovations in food labelling**
Edited by J. Albert
185 **Delivering performance in food supply chains**
Edited by C. Mena and G. Stevens
186 **Chemical deterioration and physical instability of food and beverages**
Edited by L. H. Skibsted, J. Risbo and M. L. Andersen
187 **Managing wine quality Volume 1: Viticulture and wine quality**
Edited by A. G. Reynolds
188 **Improving the safety and quality of milk Volume 1: Milk production and processing**
Edited by M. Griffiths
189 **Improving the safety and quality of milk Volume 2: Improving quality in milk products**
Edited by M. Griffiths
190 **Cereal grains: Assessing and managing quality**
Edited by C. Wrigley and I. Batey
191 **Sensory analysis for food and beverage quality control: A practical guide**
Edited by D. Kilcast
192 **Managing wine quality Volume 2: Oenology and wine quality**
Edited by A. G. Reynolds
193 **Winemaking problems solved**
Edited by C. E. Butzke
194 **Environmental assessment and management in the food industry**
Edited by U. Sonesson, J. Berlin and F. Ziegler
195 **Consumer-driven innovation in food and personal care products**
Edited by S. R. Jaeger and H. MacFie
196 **Tracing pathogens in the food chain**
Edited by S. Brul, P. M. Fratamico and T. A. McMeekin
197 **Case studies in novel food processing technologies: Innovations in processing, packaging, and predictive modelling**
Edited by C. J. Doona, K. Kustin and F. E. Feeherry
198 **Freeze-drying of pharmaceutical and food products**
T-C. Hua, B-L. Liu and H. Zhang
199 **Oxidation in foods and beverages and antioxidant applications Volume 1: Understanding mechanisms of oxidation and antioxidant activity**
Edited by E. A. Decker, R. J. Elias and D. J. McClements

200 **Oxidation in foods and beverages and antioxidant applications Volume 2: Management in different industry sectors**
Edited by E. A. Decker, R. J. Elias and D. J. McClements
201 **Protective cultures, antimicrobial metabolites and bacteriophages for food and beverage biopreservation**
Edited by C. Lacroix
202 **Separation, extraction and concentration processes in the food, beverage and nutraceutical industries**
Edited by S. S. H. Rizvi
203 **Determining mycotoxins and mycotoxigenic fungi in food and feed**
Edited by S. De Saeger
204 **Developing children's food products**
Edited by D. Kilcast and F. Angus
205 **Functional foods: Concept to product Second edition**
Edited by M. Saarela
206 **Postharvest biology and technology of tropical and subtropical fruits Volume 1: Fundamental issues**
Edited by E. M. Yahia
207 **Postharvest biology and technology of tropical and subtropical fruits Volume 2: Açai to citrus**
Edited by E. M. Yahia
208 **Postharvest biology and technology of tropical and subtropical fruits Volume 3: Cocona to mango**
Edited by E. M. Yahia
209 **Postharvest biology and technology of tropical and subtropical fruits Volume 4: Mangosteen to white sapote**
Edited by E. M. Yahia
210 **Food and beverage stability and shelf life**
Edited by D. Kilcast and P. Subramaniam
211 **Processed Meats: Improving safety, nutrition and quality**
Edited by J. P. Kerry and J. F. Kerry
212 **Food chain integrity: A holistic approach to food traceability, safety, quality and authenticity**
Edited by J. Hoorfar, K. Jordan, F. Butler and R. Prugger
213 **Improving the safety and quality of eggs and egg products Volume 1**
Edited by Y. Nys, M. Bain and F. Van Immerseel
214 **Improving the safety and quality of eggs and egg products Volume 2**
Edited by F. Van Immerseel, Y. Nys and M. Bain
215 **Animal feed contamination: Effects on livestock and food safety**
Edited by J. Fink-Gremmels
216 **Hygienic design of food factories**
Edited by J. Holah and H. L. M. Lelieveld
217 **Manley's technology of biscuits, crackers and cookies Fourth edition**
Edited by D. Manley
218 **Nanotechnology in the food, beverage and nutraceutical industries**
Edited by Q. Huang
219 **Rice quality: A guide to rice properties and analysis**
K. R. Bhattacharya
220 **Advances in meat, poultry and seafood packaging**
Edited by J. P. Kerry
221 **Reducing saturated fats in foods**
Edited by G. Talbot
222 **Handbook of food proteins**
Edited by G. O. Phillips and P. A. Williams
223 **Lifetime nutritional influences on cognition, behaviour and psychiatric illness**
Edited by D. Benton
224 **Food machinery for the production of cereal foods, snack foods and confectionery**
L-M. Cheng
225 **Alcoholic beverages: Sensory evaluation and consumer research**
Edited by J. Piggott

226 **Extrusion problems solved: Food, pet food and feed**
M. N. Riaz and G. J. Rokey
227 **Handbook of herbs and spices Second edition Volume 1**
Edited by K. V. Peter
228 **Handbook of herbs and spices Second edition Volume 2**
Edited by K. V. Peter
229 **Breadmaking: Improving quality Second edition**
Edited by S. P. Cauvain
230 **Emerging food packaging technologies: Principles and practice**
Edited by K. L. Yam and D. S. Lee
231 **Infectious disease in aquaculture: Prevention and control**
Edited by B. Austin
232 **Diet, immunity and inflammation**
Edited by P. C. Calder and P. Yaqoob
233 **Natural food additives, ingredients and flavourings**
Edited by D. Baines and R. Seal
234 **Microbial decontamination in the food industry: Novel methods and applications**
Edited by A. Demirci and M. O. Ngadi
235 **Chemical contaminants and residues in foods**
Edited by D. Schrenk
236 **Robotics and automation in the food industry: Current and future technologies**
Edited by D. G. Caldwell
237 **Fibre-rich and wholegrain foods: Improving quality**
Edited by J. A. Delcour and K. Poutanen
238 **Computer vision technology in the food and beverage industries**
Edited by D-W. Sun
239 **Encapsulation technologies and delivery systems for food ingredients and nutraceuticals**
Edited by N. Garti and D. J. McClements
240 **Case studies in food safety and authenticity**
Edited by J. Hoorfar
241 **Heat treatment for insect control: Developments and applications**
D. Hammond
242 **Advances in aquaculture hatchery technology**
Edited by G. Allan and G. Burnell
243 **Open innovation in the food and beverage industry**
Edited by M. Garcia Martinez
244 **Trends in packaging of food, beverages and other fast-moving consumer goods (FMCG)**
Edited by N. Farmer
245 **New analytical approaches for verifying the origin of food**
Edited by P. Brereton
246 **Microbial production of food ingredients, enzymes and nutraceuticals**
Edited by B. McNeil, D. Archer, I. Giavasis and L. Harvey
247 **Persistent organic pollutants and toxic metals in foods**
Edited by M. Rose and A. Fernandes
248 **Cereal grains for the food and beverage industries**
E. Arendt and E. Zannini
249 **Viruses in food and water: Risks, surveillance and control**
Edited by N. Cook
250 **Improving the safety and quality of nuts**
Edited by L. J. Harris
251 **Metabolomics in food and nutrition**
Edited by B. C. Weimer and C. Slupsky
252 **Food enrichment with omega-3 fatty acids**
Edited by C. Jacobsen, N. S. Nielsen, A. F. Horn and A-D. M. Sørensen
253 **Instrumental assessment of food sensory quality: A practical guide**
Edited by D. Kilcast
254 **Food microstructures: Microscopy, measurement and modelling**
Edited by V. J. Morris and K. Groves
255 **Handbook of food powders: Processes and properties**
Edited by B. R. Bhandari, N. Bansal, M. Zhang and P. Schuck

256 **Functional ingredients from algae for foods and nutraceuticals**
Edited by H. Domínguez
257 **Satiation, satiety and the control of food intake: Theory and practice**
Edited by J. E. Blundell and F. Bellisle
258 **Hygiene in food processing: Principles and practice Second edition**
Edited by H. L. M. Lelieveld, J. Holah and D. Napper
259 **Advances in microbial food safety Volume 1**
Edited by J. Sofos
260 **Global safety of fresh produce: A handbook of best practice, innovative commercial solutions and case studies**
Edited by J. Hoorfar
261 **Human milk biochemistry and infant formula manufacturing technology**
Edited by M. Guo
262 **High throughput screening for food safety assessment: Biosensor technologies, hyperspectral imaging and practical applications**
Edited by A. K. Bhunia, M. S. Kim and C. R. Taitt
263 **Foods, nutrients and food ingredients with authorised EU health claims: Volume 1**
Edited by M. J. Sadler
264 **Handbook of food allergen detection and control**
Edited by S. Flanagan
265 **Advances in fermented foods and beverages: Improving quality, technologies and health benefits**
Edited by W. Holzapfel
266 **Metabolomics as a tool in nutrition research**
Edited by J-L. Sébédio and L. Brennan
267 **Dietary supplements: Safety, efficacy and quality**
Edited by K. Berginc and S. Kreft
268 **Grapevine breeding programs for the wine industry: Traditional and molecular technologies**
Edited by A. G. Reynolds
269 **Handbook of antimicrobials for food safety and quality**
Edited by M. Taylor
270 **Managing and preventing obesity: Behavioural factors and dietary interventions**
Edited by T. P. Gill
271 **Electron beam pasteurization and complementary food processing technologies**
Edited by S. D. Pillai and S. Shayanfar
272 **Advances in food and beverage labelling: Information and regulations**
Edited by P. Berryman
273 **Flavour development, analysis and perception in food and beverages**
Edited by J. K. Parker, S. Elmore and L. Methven
274 **Rapid sensory profiling techniques and related methods: Applications in new product development and consumer research,**
Edited by J. Delarue, B. Lawlor and M. Rogeaux
275 **Advances in microbial food safety: Volume 2**
Edited by J. Sofos
276 **Handbook of antioxidants in food preservation**
Edited by F. Shahidi
277 **Lockhart and Wiseman's crop husbandry including grassland: Ninth edition**
H. J. S. Finch, A. M. Samuel and G. P. F. Lane
278 **Global legislation for food contact materials: Processing, storage and packaging**
Edited by J. S. Baughan
279 **Colour additives for food and beverages: Development, safety and applications**
Edited by M. Scotter
280 **A complete course in canning and related processes 14th Edition Volume 1**
Revised by S.Featherstone
281 **A complete course in canning and related processes 14th Edition Volume 2**
Revised by S.Featherstone
282 **A complete course in canning and related processes 14th Edition Volume 3**
Revised by S.Featherstone

Introduction

Heat treatment for insect control is not a new concept. The technique has been used successfully since at least 1914 in US flour mills. However, for a long time it has been regarded as a practice restricted to a few specialist operators, supplying largely the organic and environmental markets. Other methods of controlling insect infestations include low temperatures, fumigation, irradiation and microwave treatment, all of which have advantages and drawbacks. Heat treatment is now increasing in popularity as a safe, fast and effective method of insect control, replacing the fumigant gas methyl bromide, which until 2005 was the primary choice of treatment.

Methyl bromide (CH_3Br) was historically used to control insects in flour mills, food production machinery and silos, as well as in a whole host of other situations including treatment of timber and some commodities (e.g. cocoa beans, nuts, coffee and rice). Methyl bromide was highly regarded for many reasons. It has a convenient boiling point at around 4 °C, so can be stored as a liquid and applied either through a hot water vapouriser or via vapourising spray jets. It is quite good at penetrating commodities and timber, but cannot penetrate plastic sheeting.

However, methyl bromide has several disadvantages. Treatments can take upwards of 24 hours, during which time the whole building has to be evacuated in case of leakage. This makes fumigation potentially expensive, not only in terms of contractor costs but also in terms of lost production. Methyl bromide is also a highly toxic gas. Colourless and odourless, it is potentially very dangerous to operators and special precautions must be taken. It also is chemically active, often leaving behind bromide residues in food products, and can damage many products such as leather and rubber. Perhaps most importantly of all, it has been found to be an ozone depleting substance which has led to its being banned in many countries under the Montreal Protocol for Ozone Depleting Substances.

At the time of writing, phosphine gas (PH_3) fumigation remains the dominant choice for insect control as an alternative to methyl bromide. However, with treatments taking from five to 14 days, forward planning is required to accommodate the cost of the treatment area being taken out of service. Phosphine gas fumigation is also temperature dependent (not successful below 10 °C) and in some parts of the world the development of resistance or tolerance in some insect populations has been recorded, due to sub-lethal application.

As an alternative, heat treatment for insect control can be applied to buildings, food production machinery, empty bins and silos, vehicles of various kinds and even commercial airliners. New, cheap kilns made from converted transport containers have enabled the timber, pallet and packaging industries to adopt heat treatment as the main form of insect control, to comply with the requirements of ISPM 15.

Heat treatment has many advantages over methyl bromide or phosphine gas. With the right training and knowledge it is easy to use and treatments can take just a few hours compared to 24 hrs for even the most basic methyl bromide fumigation. It is also safer than other options in that people can work around and even enter into heat treatment zones, which means no forced factory shutdowns so maintenance, cleaning and production operations can often continue around heat treatment areas without hazard. Heat treatment is in many cases the only practicable choice of insect control as, unlike sprays, fogs and baits, it kills insects at all stages of the life cycle from egg to adult. A major disadvantage of heat treatment is that it is not suitable as yet for commodities. Research work is currently ongoing into this area of application, and it is likely that more applications will be developed for commodities. However, these are likely to be fixed pieces of in-line machinery rather than the mobile contractor-based services previously used for fumigation.

Successful heat treatment is both a science and an art – it requires a combination of controlled heating, temperature measurement technologies and process management to kill the target pests without causing damage to the surrounding areas. Operators must understand the basic laws of physics and thermodynamics, as well as having an understanding of insect biology and how the target pest infestation relates not only to the site of infestation, but also to the business or domestic circumstances. This book aims to identify good practice in the use of heat treatment of insect pests and its broad range of applications, particularly as part of an integrated pest management programme.

Part One

Principles

Fundamentals of heat treatment of insect pests

1

1.1 Introduction

Animals are made up of a combination of materials, including fat, bone, water, chemicals and proteins, the latter of which are largely amalgamated from amino acids. These proteins can be highly specialised, and range from structural proteins such as keratin (which makes up hair and fur) and haemoglobin (which carries oxygen and carbon dioxide in the blood) to antibodies and enzymes. Being quite fluid and complex in design, proteins are not generally heat stable and their functionality decreases sharply to the point of destruction at temperatures just outside the normal ranges of life on earth. Normal human body temperature is around 37 degrees Celsius (98.4 degrees Fahrenheit). Temperatures over 40 degrees Celsius are a potential risk to health. The reason for this is that the structures of the proteins in our bodies start to weaken and cease to function as they should.

However, humans are able to endure extreme forms of heat, such as saunas, due to the high water content of the human body. This allows perspiration, incorporating the principle of *latent heat of evaporation*. In addition, the relatively large mass of humans compared to smaller animals and insects carries benefits in terms of *specific heat capacity*. These are two basic terms in heat dynamics that should be understood when trying to carry out heat treatments for pest control. Animals can also survive at high temperatures, for example in deserts at temperatures of over 50 °C, by avoiding the heat, moving about largely at night and living under the surface of the sand by day, thereby exploiting the *insulation properties* of the sand or rocks under which they hide. If desert animals need to move about on the surface by day, they do it very quickly and by minimising surface contact as far as possible before reaching a point of heat refuge. Typical avoidance techniques include the classic ripple movement of the sidewinder snake as it moves across the hot surface, keeping only two constantly changing points of contact with the hot sand, and lizards standing on the tips of just two legs at a time to keep away from radiated heat.

Despite these avoidance techniques, a true core temperature of above 45 °C sustained for several hours will kill most complex forms of life, e.g. all vertebrates, most adult insects and other invertebrates. Some eggs or specialised diapausing or pupal states are more resistant and may require longer treatment.

Just as heat causes damage to animals, it can also cause damage to whatever the target pest is living in, from food products to machinery, cars and other transport vehicles and building structures. The purpose behind this book is to guide pest control technicians in how to kill insects and mites in food processing, storage and other environments without damaging the infrastructure (e.g. building or machinery)

Heat Treatment for Insect Control. http://dx.doi.org/10.1533/9780857097811.1.3

or product in which the insects are living. In order to do this successfully, a basic understanding of heat dynamics is required.

1.2 Heat transfer: conduction, convection and radiation

Heat flows from hot to cold, and this movement is called *energy transfer*. Energy transfer can take place in the form of *conduction*, *convection* or *radiation*

When a substance is heated up, its particles gain kinetic energy. In gases or liquids this means that the particles move around more quickly, in solids they vibrate more rapidly. If these particles are packed closely together in a dense, hard product like a metal, then as one vibrates it bumps into its neighbours, causing them to vibrate and so on down the line, causing the vibrations to move through the product. This movement of energy down the line or progressive heating is called *conduction*. The closer molecules are packed together, the quicker the process of energy transfer and the better the *conductivity*. Metals are good conductors of heat, plastic and wood less so, and items such as wool or fabrics are very poor, as each segment is linked by thin fibres separated by air, which, because it is a gas, is a good *insulator*.

Conduction cannot be achieved in a gas, for example, because the molecules or atoms are dispersed, i.e. they are not physically touching each other. The movement from A to B of a given volume of hot air (or gas) is called *convection*. Convection can be forced, as in fan heaters or water pumps, or natural as in a home radiator system based on immersion heaters and radiators. In forced convection, air or water is heated up and forced along ducts or pipes from the energy source to the cold region. Here either the hot air displaces cold air to heat the area up, or hot liquid is passed through radiators, which cause convection by heating the air around them, causing it to rise and be replaced by colder air from below.

Radiation involves the emission of *electro-magnetic waves*. Heat radiation is the emission of *infrared energy*, which is a type of radiation next to visible light in the electromagnetic spectrum. Infrared radiation is continually emitted or absorbed by matter depending on whether it is colder or hotter (respectively) than the background conditions. Radiation emission and absorption is also affected by colour, with black or dark colours being good emitters and absorbers of heat, while silver or light colours are poor absorbers and emitters because they reflect heat radiation.

Infrared cameras are built to detect this form of radiation, with image signals converted into visible spectra, normally defined using a contrast between black (cold) and white (hot). In this respect, the greater the difference in temperature between the heat emitter and the background ambient temperatures, the better the image. So, for example, a man running hard over frosty ground at night without a coat (i.e. poorly insulated) will show up much more easily than the same man walking slowly wrapped up in a coat on a warm summer evening.

1.3 Measurement of energy

The basic unit of measuring temperature is on the absolute temperature scale: the degree Kelvin. This is the same basic unit as a degree Celsius except that, instead of starting at zero degrees C (the freezing point of water at standard temperature and pressure (STP)), the Kelvin scale starts at what is known as absolute zero where there is no energy at a molecular level, i.e. −273 °C. It is important to be aware that Standard Temperature and Pressure (STP) are not even standard across the world – the International Union of Pure and Applied Chemistry (IUPAC) has STP as zero degrees C and absolute pressure of 100 kPa (14.504 psi, 0.986 atm). However, the National Institute of Standards and Technology (NIST) has STP as a temperature of 20 °C (293.15 K, 68 °F) and an absolute pressure of 101.325 kPa (14.696 psi, 1 atm).

Power is measured in watts. Power is a reflection of the rate of doing work; 1 watt is 1 joule per second. Different heaters have different power ratings e.g. a 100 kW heater can produce 100 kJoules per second. *Power = work done/time taken.* As an example, if a heater passes out 6 kJ (6000 joules) in 2 minutes its power output will be: P = W/t → 6000/120 = 50 watts.

For electricity, the simple formula is: *Power output (watts) = Volts × Amps.* So from a single phase 230 v socket at 13 amps the maximum power is 230 × 13 = 2990 watts or 2.9 kWatts, which means the heater can provide 2900 joules every second. From a 415 v three-phase system on a 16 amp circuit breaker we could get 415 × 16 × 3 = 19 920 or 19.9 kW in total, minus the power needed for any fan units (leaves most three-phase heaters rated at 18–19 kW), or as sometimes termed: kVA (W = V × A).

The potential calculations for heating something up are then complicated by the opposite powers trying to cool it down. Here the insulation properties of the building must be taken into account, or more specifically the relative cross section areas of different materials such as roof tiles, insulated walls, cladding, etc., all of which have different *insulation properties*. The temperature difference between the ambient temperature outside the heat treatment area and the slowly increasing temperature of the heat treatment area *delta t*, will also greatly influence the rate of cooling – the higher the delta t, the quicker a body will cool down.

Heat treatment will plateau when the rate of temperature increase influenced by the power of the heaters is equal to the rate of heat loss as influenced by the delta t and the insulation effort. If you are doing a job that may be influenced by weather, e.g. in winter, you can influence the delta t by starting early and aiming to peak at midday when the day is at its warmest. If you leave the job until later in the day, the inside temperature will be at its peak whilst the external temperature will be in decline, leading to an increasing delta t reading and correspondingly more heat loss. This can only be influenced then by adding more insulation. Using an infra-red spot meter or thermal imaging camera will show you where you are losing most heat. It may also be possible to re-circulate heat, which if practical will make a huge difference.

1.4 Specific heat capacity and specific heat of evaporation

The measure of how much energy a substance can store is called its *specific heat capacity* (SHC); the higher this capacity the more heat is required to increase the temperature. For example, it takes about 4200 joules to warm 1 kg of water (i.e. one litre) by 1 °C, but only around 500 joules to warm 1 kg of steel by 1 °C. Materials that need a lot of heat to warm up also store a lot of heat energy, which they release as they cool down. This is of great relevance during heat treatment, because an understanding is necessary of how the significant masses of different materials within a heat treatment area will behave. This can make the difference between success and failure for a heat treatment.

An empty, airtight building 10 m high by 5 m wide by 20 m long with a well-insulated roof will heat up far more quickly than a building of the same size full of heavy machinery. Just two cubic metres of water (e.g. stored in tanks for fire-fighting) will act as a massive energy sink, particularly if they are uncovered, as this brings into play another element of energy transfer: *specific heat of evaporation.* Water boils at 100 °C at sea level, where it can exist as both liquid and gas (water vapour/steam). In order to turn from liquid into gas, the loose bonds holding the molecules of water together as a liquid must be broken. The energy needed to boil 1 kg of material without changing its temperature is called the specific heat of evaporation, and this is different for different materials.

For water, the specific heat of evaporation is 2260 kJoules per kg (or litre). Just 4.2 kJ of energy is required to warm 10 Kg of water by 1 degree K (or C), but it takes over 500 times more energy to evaporate one litre of water than it does to heat it up by 1 °C. This has huge significance for heating up timber, commodities and in drying out buildings, e.g. after flood damage.

To calculate the energy required to heat up a material:

$$\underset{\text{(Joules)}}{\text{Energy}} = \underset{\text{(Kg)}}{\text{Mass}} \times \underset{\text{(J/Kg/K)}}{\text{Specific heat capacity}} \times \underset{\text{Delta t (K)}}{\text{Temperature change}}$$

The SHC of steel is around 500 J/kg/°C (depending on the type of steel), so to heat up 5 tonnes (5000 Kg) of steel from 20 °C to 52 °C will require 5000 × 500 × 32 = 80,000,000 joules or 80,000 kJ. Technically speaking, using a 100 kW heater, if all of the energy output from the heater were converted into heating up the 5 tonnes of steel, then the job would take:

80,000 kJ/100 kJ/sec heater output = 800 seconds or 13.3 minutes

However, not all the heat is used, and if you are using a hot air convection system then most of the air blows over the steel and out the other side, unless you have set up a recirculation system.

Energy required to evaporate a material is:

Energy (Joules)	=	Mass (Kg)	×	Specific latent heat of evaporation (J/Kg)

For example to evaporate 10 litres of water (10 kg), will require 10 kg × 2,260,000 = 22,600 kJ, once it had reached boiling point. It is therefore useful to mop up any surface water (e.g. from cleaning) before starting a heat treatment job.

1.5 Target temperatures for the heat treatment of insects

There is a lot of documented information on the efficacy of heat against insects. There is less data from field situations, which are very different to laboratory experiments. Fields (1992) provides one of the best summaries of research in this area. Studies are often based on different parameters and are geared towards grain pests, so it is difficult to extrapolate to other field situations. Another big difference is that, in the laboratory, eggs and pupae are the most difficult stages to kill. However, in the real world insects at these stages are easier to kill, as unlike larvae and adults (particularly cockroaches and bedbugs), they cannot move to avoid heat.

Broadly speaking, 50 °C for one hour will kill all life stages of most insect and mite species. However, this is assuming that the heat treatment technician has measured the temperature in the coldest part of the job, that he is measuring the actual temperature of the area in which the insect is living and that the temperature sensor is not being influenced by the hot air. To allow a reasonable margin of error, a target temperature of 52 °C for one hour in the coldest spot is suggested. For the purposes of ISPM15, the IPPC require a temperature of 56 °C be achieved for 30 minutes in the coldest part of the kiln inside the thickest piece of timber (IPPC, 2009). Table 1.1 summarises the effect of temperature on insect populations.

In hotter climates, or when treating an infestation indoors, achieving 50 °C for one hour is generally not a problem. However, in European winters, outside or in unheated buildings this can often be an issue, particularly where the starting temperature is in single figures. In this case a temperature of 47 °C for four hours is

Table 1.1 **Response of insect pests to high temperatures**

Temperature range (°C)	Effect on insects
25–30	Optimum for development
30–36	Maximum temperature for reproduction of most species
36–42	Populations die out, mobile insects seek cooler zones
42–50	Death within a day
50–60	Death within an hour
Above 60	Death within a minute

Source: Modified from Banks and Fields (1995), and Burks *et al.* (2000).

normally sufficient to kill most insects and their eggs, given the extra effect of thermal shock.

1.6 Heat treatment in practice

The following are practical tips for successful heat treatment in practice. These are discussed in more detail in the following chapters.

Be aware of energy waste and heat loss:

- Hot air should be recirculated wherever possible. This will depend on the individual job – it is not always practical in buildings, whereas in kilns, recirculation is almost essential.
- If doing multiple small jobs, plan the order of work so you can use the waste air from the job in hand to pre-heat the next job. A simple duct from the air output points on the job in hand can be fed under the sheets of the next job.
- Insulate the job as well as possible – multiple layers are better than one thick tarpaulin. If operating at the edge of your energy capacity, this is essential. If heat can be felt on the outside of the stack, the insulation is probably insufficient.
- Do not forget to insulate air or hydronic fluid input pipes as these are key areas for energy loss.
- If using a convection system, do not waste energy by heating up cold air from outside. Where possible, warm air should be drawn in from inside the building, or radiated heat from the heater unit used to pre-heat air before it enters the heater fan. Most heaters work on a delta t principle, in that their maximum output temperature is a fixed figure greater than the input temperature of air entering the unit. At an ambient temperature of 10 °C, a heater with a delta t figure of 50 °C can only produce a maximum of 60 °C out of the front of the unit, which may then have to be ducted 20 m across cold air, possibly losing another 10 °C in transmission and leaving only air at 50 °C exiting into the target stack. However, if the heater draws air in from an office lobby, for example at 20 °C, and the input ducting is insulated sufficiently that only 5 °C is lost in transmission, then the input temperature into the stack becomes 65 °C – ideal for heating up the stack contents without causing damage.

Be aware of heat distribution:

- There is no point wasting energy and heating one area up to 70 °C when other areas are struggling to make 45 °C. Energy resources and air distribution ducting should be managed to provide an even distribution of heat, and once an area has reached its target temperature for the required length of time, re-deployed to where they are required.
- At heat treatment operating temperatures, insects are unable to suddenly move several metres to get to slightly cooler areas. Most adult insects are incapable of significant directional movement above 45 °C and adult moths in particular become incapacitated very quickly after exposure to air temperatures above 50 °C.
- Identify the cold spots – the only reason heat treatments fail is because insects have not been heated up to a lethal temperature for long enough. This is either due to the operator in charge deliberately and misguidedly placing temperature sensors in favourable locations in order to 'complete' the task as quickly as possible, or failing to identify cold spots or refuges where insects can survive. These may include thick residues of stock, harbourages in insulated materials such as clothing or blankets, or for more mobile creatures such as

cockroaches or bedbugs there may be escape routes along cable conduits, into wall cavities or down drains.

- The location of temperature sensors is crucial to the job in hand – they should be positioned to tell the operator in charge where the energy is going (i.e. the distribution of hot air) and should be placed in the coldest, highest-risk areas to indicate the effect that heat is having on the target areas. Heat will penetrate all areas eventually – even a one tonne tote bag full of peanuts can be heated up to the core to achieve 50 °C (this takes three days). Temperature sensors should also be used to protect sensitive areas and machinery from damage. The operator in charge should agree maximum temperature exposures of risk items with the client, then place temperature sensors in locations agreed with the client, so that in the event of a claim for damage, the operator in charge is able to show he has worked within the agreed parameters.
- Possible escape routes should be identified at the survey stage and either physically blocked or treated with insecticide so that insects running over them are killed as they try to escape.

1.7 Summary

Understanding the physics of a heat treatment is important to achieving the desired result. However, performing a lot of complicated mathematical calculations would normally be impractical given the time constraints each job requires, particularly given that the price quoted may or may not be accepted by a client. The simple solution here is to make sure you have more than enough resources to hand to do the job, and the more effort put into insulating the area efficiently (i.e. controlling potential heat loss), the better your chances of success.

References and further reading

Banks, J. and Fields, P. (1995), 'Physical methods for insect control in stored-grain ecosystems', in Jayas, D. *et al.* (Eds.), *Stored-grain ecosystems*, Marcel Dekker Inc., New York.

Bell, C. (2014), 'Pest control of stored food products: insects and mites', in Lelieveld, H., Holah, J. and Napper, D. (Eds.), *Hygiene in food processing: principles and practice* (Second edition), Woodhead Publishing Limited, Cambridge, UK.

Burkes, C., Johnson, J., Maier, D. and Heaps, J. (2000), 'Temperature', in Subramanyam, B. and Hagstrum, D. (Eds.), *Alternatives to pesticides in stored-product IPM*, Kluwer Academic Publishers, The Netherlands.

Fields, P. (1992), 'The control of stored-product insects and mites with extreme temperatures', *J. Stored Prod. Res.* 28(2): 89–118.

Heaps, J. (Ed.) (2006), *Insect management for food storage and processing* (Second edition), AACC Publishers, USA.

Hill, D. (2002), *Pests of stored foodstuffs and their control*, Springer Publishers, The Netherlands.

IPPC (2009), International Standards for Phytosanitary Measures, Revision of ISPM No. 15 'Regulation of Wood Packaging Material in International Trade'.

Kloosterman, L. and Mager, K. (2014), 'Pest control in food businesses: an introduction', in Lelieveld, H., Holah, J. and Napper, D. (Eds.), *Hygiene in food processing: principles and practice* (Second edition), Woodhead Publishing Limited, Cambridge, UK.

Morland, D. and Hedoes, S. (Eds.) (2004), *The Mallis handbook of pest control* (Ninth edition), GIE Media Inc., USA.

Subramanvam, B. (Ed.) (1995), *Integrated management of insects in stored products*, CRC Press, Boca Raton, USA.

Operational planning, best practice and cost issues in heat treatment of insect pests

2

2.1 Introduction

Heat treatment can seem daunting because of the amount of equipment required compared to more conventional pest control operations. However, the size and complexity of the equipment is deceptive in that it is mostly quick and easy to set up and pack away. In many cases heating can commence within an hour of the team arriving on site.

Like most other forms of pest control, treatment will start with a detailed survey. This is an important stage in the treatment process. The survey determines the amount of equipment required, fuel, staff, risks and health and safety assessments, client liaison and client responsibilities. Detailed temperature logging takes place during the treatment, so that a report can be prepared afterwards and sent to the client. Reports usually include heat treatment graphs, an assessment of the success of the treatment and future recommendations.

Before a survey is conducted, the operators should determine which types of heat treatment they are able to offer, and the amount and type of equipment required to carry out the job. Too large a fan will do nothing more than blow relatively cool air into the stack and rapidly out the other side before it has had a chance to give up its energy. Conversely, too small a fan will not push air through the heat treatment area, causing hot air to back up into the heater and consequently over-heating and tripping the safety thermostats. Big fans, for example, are useful for building jobs where maximum air turbulence enhances heat distribution and penetration. Glycol or steam radiator systems are notoriously cumbersome and time consuming to set up, but may be the only practicable way of treating very large, complex structures. Forced air convection systems are cheaper in terms of capital costs and very quick and easy to use, but less energy efficient. With experience the heat treatment technician will come to understand how his heaters behave in different circumstances and how to account for variables, such as weather, on treatments.

The other major consideration concerning equipment is transport. In the UK and parts of Europe it is easy to move heating kits around in trailers. However, trailers are demanding to tow over long distances. Some companies opt to use box vans. Attention must be paid to safe loading and unloading. Tail lifts, hoists and ramps have all been used successfully to load and unload heavy heating equipment.

It is worth noting the importance of technician training. This book is designed to be part of a training course and may be used as a framework for a course syllabus,

Heat Treatment for Insect Control. http://dx.doi.org/10.1533/9780857097811.1.11

but training should include a large practical element. At Thermokil Ltd, as suppliers of heat treatment equipment and technology, the author and his colleagues saw right from the start that there was no point in just selling equipment without teaching people how to use it safely and effectively. Whilst there are as yet no formal training requirements for heat treatment technicians, the UK British Pest Control Association runs a heat treatment course. Thermokil Ltd also provides formal training with each heat treatment kit it sells, ranging from three to five day courses. Pest control companies should treat the training and use of heat treatment technicians as if they were fumigation technicians. You should no more put an untrained person in charge of a heat treatment than you would put a non-qualified person in charge of a fumigation.

The stages of heat treatment are as follows:

1. Survey of job.
2. Execution of job.
3. Final report.

2.2 The survey

The survey is an essential stage in heat treatment operations, which should be conducted with due attention. The following points should be covered:

- Where is the infestation and how did it get there?
- What other supplementary insect control measures will be required, and what health and safety and risks are associated with them?
- Which areas specifically require heat treatment? Which areas will be treated with conventional pesticides that will leave a residual action long after the heat treatment team have left? Will the whole building be heated or would spot treating specific areas or machinery be more efficient?
- Having determined the areas to heat treat the surveyor needs to know the rough mass of the equipment, the volume of the area, and how he is going to isolate and contain the heat within the designated area. Does he have enough sheets or need to buy in disposable polythene?
- Will heat loss be significant? If so, how much insulation will be required and where? Mistakes are often made at this stage. For example, free standing metal silos need insulation, and buildings in Mediterranean countries are often poorly insulated. Flour mills in cold areas like central and northern Europe are normally well insulated and need little preparation.
- Having determined the heat treatment area, are there any items within that area that could be damaged by heat? Never expose anything other than bare metal or wooden pallets to temperatures over 70 °C. Exposures above this run a very high risk of damage. Core temperatures should ideally not exceed 60 °C, and there is no need to go above this because insects are very quickly killed at this temperature. If one part of the heat treatment job is at this temperature whilst other areas struggle to reach lethal temperature then the operator should take measures to re-balance heat distribution.

The other major function of a survey is of course to perform a risk assessment. Table 2.1 outlines risks associated with heat treatments between the range of 50 °C and 70 °C (higher temperatures are deemed too risky by the author and other heat treatment professionals).

Table 2.1 Risks associated with heat treatments

Material	Risk	Notes
Electrical equipment Switched on Switched off	High Low	The basic physical material of most electrical components is heat resistant to 70 °C, circuit boards, solder wiring and silicon chips are not going to melt. Indeed all airline equipment is designed to work at 65 °C or more, so that it can safely operate in deserts with margin to spare. However once you put a current on equipment two things can happen. First, the natural heating from the current plus the additional heating of the insect control job can cause local overheating or damage; and secondly, a lot of modern computers and microchips rely on minute changes in micro voltages that can then be enhanced to cause major changes elsewhere in the process. During heating the electrical resistance of different parts of the machinery will change, thus changing the current, which can give false results and cause computer malfunction. So the golden rule is always switch off all electrical equipment. From a purely practical and health and safety point of view, all physical machinery should be electrically isolated and locked off anyway.
Glues	High	There are a huge range of glues in use, and some are not very heat stable. If in doubt either consult the technical specification for the material or get the QA department to test the component in an oven. A household electric oven is sufficient for this.
Paintings	High	Heat treatments should not be carried out without specialist conservator advice.
Plastics	Moderate	Some lightweight plastics and polythene can be effected by temperatures as low as 60 °C or even less and may crinkle or melt. Do not directly heat lightweight or flexible plastic.
Conveyor belts	Low to moderate	Most food production conveyors are quite resilient, however some of the lighter weight ones may be at risk if heat is applied directly. Keep below 60 °C.
Woollen clothing	Moderate. More research needed.	No obvious damage has been observed from heat treating between 47 and 55 °C. However, fibres deep in the material could be weakened or damaged. Whilst you would never wash woollen clothing at 55 °C, it could be that the physical action of the washing process causes damage. Be careful and if in doubt do a test first in an oven.

(*Continued overleaf*)

Table 2.1 Continued

Material	Risk	Notes
Floor tiles (vinyl)	Moderate	The main area for concern here comes from adhesives, which could weaken or contract, curling or lifting the tile. Try to avoid direct heat onto flooring. Again there are huge variations and unknowns; if in doubt, test first.
Floor tiles (ceramic)	Small to moderate	There is zero risk of a ceramic tile cracking from exposure to 70 °C heat. However, if a metal structure is tightly integrated into a ceramic tiled floor, expansion in the metal work could cause cracking in the tiles.
Wooden floors and buildings	Moderate	As with all wood, the risk comes from water transfer, either in or out, or uneven water movement.
Pneumatic air lines	Low	During a heat treatment, pneumatic pressurised air lines operating valves or slides, etc., might go slightly soft and, if under pressure, could be blown off. Always turn off all compressed air supplies.
Pressure vessels	Moderate	Always remove anything sealed or pressurised. Heat will cause expansion of air and this can even blow tins of food. Fire extinguishers and halon gas dispensers should be removed or protected.
Smoke alarms	Moderate	There can be considerable air movement during some heat treatments that can disturb dust on tops of wardrobes, etc., which smoke alarms interpret as smoke. These should therefore be covered, and fire brigades warned of forthcoming treatments to prevent false alarms.
Fire detection and control systems	High	The biggest issue can be with sprinklers. Whether you choose to remove heads, risk gently heating them but turn off the main stop tap, or aim to cover and protect them, will depend on circumstances. As a heat treatment contractor it is best to leave this to the client to organise. Fire control systems will almost certainly need to be turned off if insurance will allow it. Heat treatment jobs should never be left unmanned, so there will always be a manned fire patrol to cover, and it just takes pressing a few buttons to re-activate a fire control system in case of emergency.

2.3 Execution of the heat treatment

The following points can be used as a checklist when carrying out heat treatments:

2.3.1 Before heat treatment

1. Select an operator in charge to coordinate the heat treatment.
2. The survey will have identified specific areas to be heated and survey notes or plans should be provided. These must be reconciled with reality and discussed with the site contact.
3. The survey will have identified heat sensitive items including structures and supports, such as roofs. If protection or engineering assurances cannot be agreed, then do not continue with the heat treatment.
4. Identify and develop measures for protecting heat-sensitive equipment within the facility. Contact the manufacturer if in doubt as to which equipment is heat-sensitive.
5. Identify measures to prevent either insect escape from heat treatment areas or re-infestation routes. This may include use of residual insecticide, proofing and other insect control or integrated pest management measures.
6. Determine the air movement plan, circulation equipment, heater and fan placements, and type and number of heating units and fans needed.
7. Identify energy sources, and locate an electrical panel to plug in fans and extension cords, to spread out the electrical load.
8. Establish a fire protection plan. Check the insurance carrier for coverage of any damage to the structure or equipment.
9. Repair damaged doors, windows and other openings that would allow heat to escape. This is not a major issue when using forced-air gas heaters. Seal unwanted gaps around heating duct inputs and ensure windows can be opened in appropriate areas if using forced convection systems to aid with directing airflow. Eliminate major drafts from unheated areas.
10. Notify the client's corporate safety, engineering and management personnel of your intention to conduct a heat treatment.
11. Ask the client to notify a local fire station if appropriate in case of false alarms.
12. Ensure you have agreed staff rotas for long jobs such as space treatments, planning for sleep and rest periods.
13. Use 4–6 mm polyethylene sheets to seal off exhaust fans, dust collectors or air handling systems that exhaust to the outside.
14. Remove heat-sensitive products or raw materials from the area. Examples are vitamins, shortenings, sugar and some packaging materials. Many products are sensitive to the high temperatures used during heat treatments.
15. Empty storage structures like bins or silos and run all machinery empty. Grain or grain products (flour, etc.) are good insulators, so heat will not penetrate into the bulk and insects may survive if the stored product is infested. If the stored product is insect-free and it is necessary to keep some product *in situ* for start-up or production reasons, bin entry and exit points must be sealed so that insects do not migrate into these storage structures. Check that high temperatures will not alter the quality or usability of stored products.
16. Empty rubbish bins and vacuum cleaners. Bagged, bulk raw or processed products should be placed in a trailer and fumigated with phosphine to kill residual infestations.

17. Remove pressurised containers and cylinders from the heated area. Label fire extinguishers with the proper location for emergency "near-by use" during heat treatments.
18. Remove or empty beverage vending machines.
19. Where possible, remove electronic equipment. Unplug equipment that cannot be removed. Back up computer programs. An experiment run at the Kansas State University heat workshop in 1999 showed no adverse effects on computers after they were subjected to a 34-hour heat treatment.
20. Sprinkler systems need to be protected. Agree sprinkler head temperature ratings and max temperatures with the client, measuring these and agreeing locations of safety temperature sensors with client in advance. If heads activate at lower temperatures, either drain the system, remove them and post a fireguard during the treatment, or isolate water supply and cover heads by putting a damp tissue in a plastic cup and taping over the head. Place a temperature sensor in the highest risk areas. Cups can be topped up with cold water during treatments.
21. Turn off older sodium or mercury vapour lights during heat treatment. Check with engineering staff or the supplier regarding heat tolerance of these lights. Identify alternative lighting plans to minimise plant power usage.
22. Check bearing and belt types and loosen where necessary.
23. Check lubricant type and reservoirs, and provide for expansion during heating.
24. Identify plastic-type material, including PVC piping and Tygon tubes, and monitor these for possible damage during heat treatment. Check pneumatic line plastic connectors for any adverse heat-related effects. Turn off all pneumatic air supplies.
25. Double check temperature limitations on all high risk equipment such as electronic controllers, small computers, or photoelectric sensors. The best information source is the equipment supplier, but the chief engineer must accept responsibility for either defining maximum temperature exposure levels for each piece of equipment or removing it to a safe cold zone. Sensitive equipment can be protected either by placing it in a cool zone during the heat treatment or by covering with damp cloths that are replenished during the course of a treatment. Develop floor-by-floor and area-by-area checklists for protection of sensitive equipment within heat treatment zones.
26. Take precautions about magnets that may be deactivated as a result of exposure to high temperatures (50–60 °C). Contact the manufacturer for maximum temperatures.
27. Establish an employee safety plan that covers heat illness warning signs; use of the buddy system (people working in teams of two); tips on proper clothing, drinking, eating, a heat stress first aid room outside heated area; a first aid kit; emergency phone numbers, and employee heat tolerances.
28. Identify and provide appropriate personal protective equipment (PPE) such as bump caps with cloth linings and cloth gloves. It is advisable to wear light, loose fitting clothing.
29. No metal or glass such as buttons and glasses, which are good heat conductors, should be in direct contact with skin, particularly necklaces and medallions.
30. Establish a temperature-monitoring plan, including key locations to be monitored with remote temperature logging devices.
31. Identify all areas adjacent to the heated areas. Spray surfaces, especially floor–wall junctions and doorways, with a residual insecticide to prevent insect migration to unheated areas.
32. If bio-assays are to be used, pre-determine locations on the plant layout for the placement of test insects. Cages should contain an insect that is a problem within the facility, and temperatures should be measured near the test insects. It is important to use the most heat-tolerant stage of the insect. The best procedure is to expose all stages (eggs, young larvae, old larvae, pupae, and adults) of the insect species.

33. Thoroughly clean accessible equipment, leaving no more than 1 cm thickness of food products. Proper cleaning is essential to an effective heat treatment because grain or stored products are poor heat conductors.
34. Elevator and conveyor boots are good sources of insect populations. Areas under the elevator buckets are good harbourage points for insects because broken, damaged grain becomes trapped or encrusted in these areas. Opening the boots of bucket elevators and conveyors and directing fans to these areas helps kill residual insect populations. Sometimes elevators may be run for a few hours before heat treatment completion in space treatments so that all areas of elevator (belts, cups, and screw conveyor) are exposed to lethal temperatures.
35. Identify the chief person responsible for both isolating equipment and operation of automatic slides, etc., that may be needed to be open or closed to facilitate air movement patterns. Machinery must be isolated and locked off during spot treatments.

2.3.2 During heat treatment

1. Before heaters are turned on, walk through the facility with the heat treatment team. Determine whether the level of cleaning is adequate and ensure that all the critical items have been removed from the facility.
2. Measure the temperature from as many locations as possible within the facility to identify cool (below 50 °C) and overheated areas (above 60 °C), and record this information using appropriate data logging equipment (e.g. thermolog system).
3. If using internal heaters such as glycol/steam radiators or electric heaters, numerous fans capable of withstanding 50 to 60 °C should be used to circulate hot air within the facility, and force it into corners. Appropriate circuit breakers should be used on the temporary electrical panels for fans so they do not trip.
4. Check areas near test insects regularly. Remember that insects exposed to sudden heat shock appear dead but may revive if removed from heated areas. Insect test cages with adults removed during heat treatment should be kept at room conditions for 24 hours before insect mortality is assessed. All pre-adult stages should be reared to adulthood for mortality assessments.
5. Designate an office area as a heat treatment command centre. It should contain a phone, first aid kit, temperature log sheets, fluids (water or other hydrating beverages), and emergency phone numbers.

2.3.3 After heat treatment

1. Before turning off the heat, always thoroughly inspect any test insects, known infestations and risk areas to check for live insects. If any are found, extend the treatment period as appropriate.
2. Providing all insects are dead, discontinue heating after the desired exposure time and temperature have been achieved. Keep fans running after shutting down the heaters to facilitate cooling. In the case of forced-air heaters, blowers may be left running after the burner is shut down, forcing cool air from outside into the facility.
3. Uncover roofs and wall vents, air intakes, and other openings for exhaust air. Open screened windows.
4. Turn on plant power once the temperature has cooled to less than 43 °C.
5. Recover test insects and temperature-sensing equipment or charts.
6. Start the exhaust fans in heated areas. Monitor temperatures during the cool-down period until they are below 40 °C.

7. Replace fire extinguishers at proper locations, and return the plant to normal fire-protection standards. If the sprinkler system was drained, check each sprinkler head before activation and refill slowly.
8. Remove portable power or heater equipment and begin re-assembly of plant equipment to prepare for normal operation.
9. Remove all sealing equipment and complete the post-treatment clean-up. Flush the initial food material within the processing system for about 10 to 20 minutes and dispose of it as trash. Sometimes a very large number of dead insects exit processing equipment in the initial flush. Check flushed material and record information on the types and number of insects present.

2.4 Heat treatment report

1. The heat treatment team should review treatment activity and effectiveness, and list suggestions for future improvements. Subsequent heat treatments should be more effective than the first one. Modifications are often needed due to the unique aspects of the structure and training of personnel.
2. Prepare a detailed post heat-treatment report that gives details of the treatment, showing all graphs and temperature logging locations, and assess the success or otherwise of the treatment. This report should serve as a baseline for future heat treatments.
3. Any adverse effects observed should be investigated and a plan developed to prevent such occurrences in the future.

2.5 The economics of heat treatment

The price of heat treatment is often now lower than that of the equivalent methyl bromide operation. In terms of basic economics, both treatments generally require two or more people; however, heat treatment has significant advantages over methyl bromide:

- Sheeting and preparation times are much shorter, as the treatment area does not need to be sealed to prevent gas escaping. Preparation for heat treating a small flour mill might take three or four hours, versus 2–3 days for methyl bromide.
- Raw material consumable costs are energy for heating (generally oil, but occasionally electricity) and of course methyl bromide for fumigation. In comparison, a 4000 m^3 flour mill will take £350–£400 worth of fuel oil to treat with heat, and methyl bromide gas costs at 32 g/m^3 and £4 (€5) per kg would be £512, even without the inevitable need for topping up due to leakage.

The big difference, however, is seen in the capital costs of the equipment, and the capital element per job is a function of the number of jobs carried out during the depreciation period of the equipment. A typical methyl bromide application kit for such a mill, either with cold gas hoses and jets or warm gas vapourisers, would be about £2000, whereas a simple heat treatment kit using convected air heat exchange units for the same job would be around £20,000 to £30,000. This might sound a lot of money, but if a heat treatment kit costing £30,000 is used just once a week for five years then that is just £115 per job.

As a comparative example, our 4000 m^3 small flour mill in Europe in summer would need 400 kW of heating capacity typically, which together with distribution ducting and manifolds etc. would cost around £25,000 at 2013 prices. If labour is priced at 400 GBP per man per day, the cost to the client (assuming a 100% mark up by the contractor on consumable prices) would be as follows:

Methyl bromide
Labour – 2 men for 5 days (3 to prepare, 2 to fumigate and de-gas) – £4000
Overnight charge – £500
Gas costs – 128 kg MB plus 25% top up = 160 kg x £8.00 = £1280
Equipment depreciation assuming weekly use over five years – £2000/260 = £7.70
Sundries – tape, single use polythene sheets, etc., for gas tight sealing – £400
Total charge to customer – £6187

Heat treatment
Labour – 2 men for 2 days (heat on almost immediately, minor blocking of any particularly leaky doors, setting up temperature sensors, etc., done whilst heat is on) = £1600
Overnight charge – £500
Oil cost – *c.*400 litres red diesel at 90p per litre × 100% = £720
Equipment depreciation – £96
Sundries – £200 (much less sealing required than for methyl bromide)
Total charge to customer – £3116

If the heat treatment contractor is using a large hydroponic heat treatment system (typically costing £65,000) the figures should be adjusted. For example, the capital costs might be £406 per job, but oil costs should be lower to reflect the higher efficiency.

Pricing for jobs has to take into account:

- Labour costs for preparation and monitoring – as equipment tends to be large and heavy, at least two men are generally required for setting up and preparing the job. However, for ongoing monitoring only one man is required to keep an eye on fuel and to make adjustments to air distribution as necessary. Heat treatments should never be left unattended, for reasons of health and safety and also to avoid loss of efficiency through equipment failures going unnoticed. Should direct burn propane methods be used, a two man fire watch is required 100% of the time because of the potential dangers from fire and fumes.
- Capital costs of equipment – heat treatment equipment is not cheap, particularly glycol systems which can cost over £100,000. The amount attributed to each job should be a function of how often the equipment is used versus the depreciation schedule attached to it by the accountants.
- Fuel and electricity have to be taken into account – for small jobs like silos or sheeted over machinery or furniture, this will be quite insignificant.
- Sundries such as sheeting and other support equipment can be charged (with a mark-up) if consumed or depreciated along with the equipment schedule.
- Business overheads and profit margins will also need to be considered when setting prices.

However you decide to price jobs, the survey is most important and time should be taken to assess the following points:

- Is the job feasible? The sorts of things that can stop a heat treatment in its tracks include:
  (i) Lack of big enough heaters for size of job.
  (ii) Too much water in the heat treatment area causing cooling.

- **(iii)** Ambient temperature too cold.
- **(iv)** Thick piles of clothing or heavy food or commodity residues present, which may prevent adequate heat penetration in the treatment time available.
- **(v)** Client activity around the treatment area during proposed treatment time interfering with heat treatment process.

- Area(s) to be treated and possible routes of re-infestation, control of which should be assessed and priced up alongside the heat treatment work.
- How long will the preparation time take?
- What heating, ducting and fan resources are required?
- What fuel is required and can refills be obtained when needed? Inadequate fuel supplies may seriously impair a heat treatment, as the treatment area may cool down during delays while fuel is found.
- Labour requirements.
- Sundry resources.
- Health and safety assessment and planning for water supplies, rest breaks, etc., for workers.

2.6 Summary

All heat treatments need to be planned. The planning stage might seem daunting at first, but in time it becomes second nature, whether formally documented or not. Handling of heavy duty equipment can easily become routine, though it is true that some systems are easier to use than others. Finally, it should be remembered that insects are only killed in the final hour or two of the treatment. The final hours of a heat treatment are key to its success.

References and further reading

Bell, C. (2014), 'Pest control of stored food products: insects and mites', in Lelieveld, H., Holah, J. and Napper, D. (Eds.), *Hygiene in food processing: principles and practice* (Second edition), Woodhead Publishing Limited, Cambridge, UK.

Heaps, J. (Ed.) (2006), *Insect management for food storage and processing* (Second edition), AACC Publishers, USA.

Hill, D. (2002), *Pests of stored foodstuffs and their control*, Springer Publishers, The Netherlands.

Kloosterman, L. and Mager, K. (2014), 'Pest control in food businesses: an introduction', in Lelieveld, H., Holah, J. and Napper, D. (Eds.), *Hygiene in food processing: principles and practice* (Second edition), Woodhead Publishing Limited, Cambridge, UK.

Morland, D. and Hedoes, S. (Eds.) (2004), *The Mallis handbook of pest control* (Ninth edition), GIE Media Inc., USA.

Subramanvam, B. (Ed.) (1995), *Integrated management of insects in stored products*, CRC Press, Boca Raton, USA.

Heat treatment as part of an integrated pest management programme

3

3.1 Introduction

A major failing in pest control has been the search for the perfect pest control solution, be it the perfect mouse trap, the best insecticide, or complete eradiation of pests through fumigation. Heat treatment is not a panacea but should be used as in conjunction with good cleaning regimes and as part of an integrated pest management programme. The focus of this chapter is to attempt to define what part heat treatment should play in an integrated pest management programme.

3.2 Integrated pest management programmes

There have been numerous definitions of an integrated pest management programme. Broadly speaking it requires that the client, the pest control contractor and any other interested parties work together to achieve as close to zero infestation levels in the designated area. This should be done by working to a management plan and using all proactive pest prevention techniques, supported (where these are inadequate) by reactive pest control measures designed not to control infestations but to eliminate infestations completely.

Fumigation techniques typically only kill 90–99% of infestation Many fogging agents and pheromone disruption products fail to kill eggs, buried larvae or pupae. The high levels of gas concentration required to kill insect eggs creates a large diffusion gradient which leads to increased rate of gas loss to atmosphere or adjacent areas, resulting in an incomplete treatment. Only extremes of heat, cold, fumigation and radiation can kill 100% of all life stages of insect infestations, and only a well thought out integrated pest management programme can create the conditions that allow these measures to succeed and ensure proactive measures are taken to eliminate or reduce as far as possible the risks of re-infestation. Integrated pest management programmes are particularly necessary for the control and elimination of stored product insect (SPI) infestations in high risk premises such as bakeries, flour mills and confectionary plants.

Heat Treatment for Insect Control. http://dx.doi.org/10.1533/9780857097811.1.21

3.3 Proactive pest management: exclusion, restriction and destruction

Every pest infestation is different, and it is a feature of pest control technicians' work that no two days are the same – each new infestation problem presents new challenges. Pest control technicians have to select from a range of products, application equipment and mechanical control devices to eliminate the current problem, without endangering the safety of consumers, client workers or themselves.

The goal of any manager responsible for pest control on a site, whether food orientated or otherwise, is to achieve a consistently low or ideally zero level of infestation. The only way this will be achieved is if proactive measures are taken to prevent infestation along with reactive measures to control any existing infestation. This is a general rule of pest management, whether dealing with insects, birds, rodents or other pests.

These basic principles of proactive pest management have been summarised in many ways, one of the best being the E.R.D. principle:

E: Exclusion
R: Restriction
D: Destruction (Physical or Chemical).

Although the E.R.D. principle can apply to a range of pest management situations, for insect control associated with heat treatment the following points apply.

- E: Exclusion is about keeping out infestation, thus preventing re-infestation of treated areas. This may be physical proofing of windows with fly screens to prevent ingress by free-flying insects such as the Indian meal moth (*Plodia interpunctella*) or cigarette beetle (*Lasioderma serricorne*); meshing drains to prevent oriental cockroach (*Blatta orientalis*) intrusion from sewers; or only taking ingredients from approved suppliers who are regularly audited to show they are not exporting infestation to their customers.
- R: Restriction is about making life as difficult as possible for any pests that do gain access to the premises by eliminating breeding harbourages. Examples of this are cleaning away undisturbed food residues; and storing ingredients and raw materials in pest free vessels and areas under pest averse conditions – keeping them cool and dry. For some major pests like booklice (*Pscocids*), fruit flies and mites, elimination of moisture alone can achieve complete control. Heat with forced ventilation systems is a powerful lethal combination of heat and drying for these types of insect and acarid pests. Also included under this restriction is cleaning, stock rotation, stacking, machine and plant design and physical engineering solutions.
- D: Destruction (physical) for insects encompasses the full range of heat treatment options, as well as cold treatment options, physical break up of insects in the production process (e.g. entoleters found in flour mills), physical removal of insects by cleaning, or even (in the case of bed bugs) "harvesting" of the bulk of an infestation with detailed hoovering prior to, or even in some cases instead of, heat treatment. The vacuum cleaner is a powerful and under used tool in insect control from stored product insects to bed bugs and fleas. Many serious cat and dog flea infestations in the home are the result of under-vacuumed carpets.
- D: Destruction (chemical) refers to the full range of chemical solutions: fumigants, sprays and dusts of various kinds, fogs and smokes, etc.

These basic E.R.D. principles can be applied to most pest management situations, not just insects, and indeed the principles are useful in the management of microbial infections in hospitals and food factories. With regard to stored product insect and mite infestations, as can be seen from the above principle, there are a whole range of measures that can be taken.

The first stage in a pest management programme is the riddance treatment – for insects this may typically, in a post-methyl bromide world, be a space heat treatment. It may equally be a large-scale clean down combined with numerous "spot" heat treatments of production lines and empty bins and silos. One major UK milky drink supplier has employed this technique very successfully. It may be a sulfuryl fluoride treatment or even in hot dry conditions and space a phosphine treatment. These treatments may well be enhanced by residual insecticidal spray treatments to the building fabric to kill insects flushed out by heat or fumigation and as barriers against re-infestation. Wall cavities and electrical conduits can be treated at this stage with diatomaceous earth or other suitable insecticidal dusting powders.

3.4 Exclusion techniques

The next action is to ensure there is no re-infestation. Exclusion covers a whole range of issues such as proofing, as mentioned earlier, installing fly screens and insect proof doors to stop pest moth such as Indian meal moth (*Plodia interpunctella*) from entering the building, or to stop the mill moth re-infesting from outside the fumigation area where they may survive in wheatfeed outloading bays, spillages on roof areas or leaking in dust collector or explosion vents which would have been sealed from inside.

In the milling and warehouse scenario re-infestation can take place on pallets, infested materials coming in or through pre-fumigation infested returned products that are being fed back into the system early in the mill run (as was common in the days of annual fumigations). In bakeries, these infested flour mills were responsible for infesting numerous bakery production lines with mill moth, flour beetles of various kinds (e.g. *Trobolium* spp. and *Gnathoscerus cornutus*) and red rust and flat grain beetle (*Cryptolestes* spp.).

Many bakeries responded to this with the introduction of in-line sifters to sieve flour before it went on to production, then the overtails analysis from the sieves (i.e. the ongoing recording of larger particles going over the sieve meshes rather than through) allowed them to identify offending flour suppliers with evidence, and to select from those able to supply insect-free flour all year round. By applying the exclusion principle to their sites, the bakeries enforced change on the milling industry.

It is a common misconception that, in mills and bakeries, insects breed in the main flow of flour, or other food product; however, this is not the case. Insects need security and peace and quiet in order to breed successfully. This means they need quiet undisturbed residues of food or other breeding materials, which by definition will be in those areas unlikely to be reached by the cleaning team on a regular basis.

In cigarette factories this may mean the tobacco beetle (*Lasioderma serricorne*) breeding in an old packet of cigarettes under a packing machine or in tobacco deposits in a cigarette making machine. In flour mills, the flour beetle (*Tribolium* spp.) may breed behind the feed-gates in roller mills while the flat grain beetle (*Cryptolestes* spp.)

may breed in the cracks and joins of wooden sieve frames in plan sifters. The mill moth (*Ephestia kuehniella*) thrives in the finished flour in the free spaces at the tops of flour silos and in dust collectors where free-flying adults from anywhere in the mill process get sucked into the dust extraction meshes. Here they settle and lay their eggs in areas out of the main air and product flow. The one thing all these areas have in common, along with stored food, is that they will remain undisturbed by cleaners and can breed in peace.

3.5 Restriction techniques

Restriction is all about making this breeding process as difficult as possible. It encompasses detailed cleaning programmes inside machinery, which by definition requires the co-operation of engineers to allow safe access in machinery and machine re-assembling after cleaning. It is about buying new machinery which is designed with minimal dead spaces inside. Designers are now much better at producing machinery without the internal quiet areas where insects can breed, and with flow rates through high volume machinery now much faster, this is an improving situation.

One of the most effective means of controlling insects in the fabric of the building is by tackling dust control. The free release of flour or tobacco dust into the air from leaking machinery or improperly exhausted machinery leaves layers of food product on all surfaces. This then requires daily cleaning, incurring costs, but it is often not possible or practical to clean overhead girders and other high surfaces regularly, leaving undisturbed residues free for insects to breed in. Significant levels of dust also represent a major potential explosion hazard, and even if the air is not regularly thick with dust, deposits on high levels can become a major hazard when disturbed by cleaners, engineers, birds or even pest control biologists going about their inspections. Capital expenditure on good dust control is highly economical in the long-run.

Pheromone disruption and biological control programmes can have a control function, for example when used against large infestations in buildings. Used in conjunction with heat treatment or other measures, pheromone disruption and biological control programmes have an important role to play. Most species of moth, and some free flying beetles, find their mates by smell. Normally the male can smell a female from her pheromone secretions from great distances, up to a mile in some jungle species. Developments have been made in synthesising some of these pheromones and they can be used in conjunction with trapping and monitoring programmes to aid in the estimation of levels of moth infestation. If, however, an area is saturated with high levels of pheromone, the males cannot discern where the female is, and therefore reproduction is severely disrupted and will eventually eliminate the population unless or until re-infestation occurs from outside.

Restriction does have inherent issues however – the above measures often incur costs and cause disruption to production. Many modern factories run twenty-four hours a day, seven days a week, and the full implementation of proactive control measures may not be possible, practical or cost effective. This means insect levels will build up, often in those inaccessible areas where cleaning and pest control staff cannot gain access readily. Often, too, these areas are also inaccessible to food supplier auditors who almost always visit factories during normal working hours, are often not trained pest controllers, and who rely on either sightings of insects or the pest control biologists' reports in their pest control file. This can result in infestation problems going undetected.

3.6 Destruction techniques

Destruction is important, but for insects in particular it is important that it is the right kind of destruction. Chemical fogging, misting or smoke treatments are very effective at killing exposed adult insects. The dead flies and adult moths which are found after a treatment suggest that the job is complete, however it is not a true indication that the infestation has been entirely eradicated. Almost all insects that go through the complete metamorphosis life cycle of egg, larva, pupa, adult, spend most of their lives in the first three stages. The adult stage is purely required to find a mate and allow reproduction. These adult moths that are seen lying on the ground after a fogging treatment have mostly mated and laid their eggs. The real infestation continues inside the food production machinery or in stored food and food storage areas such as bins and silos, out of the way of the fog. In many countries it is illegal to allow non-fumigant pesticides to come into contact with food contact surfaces without then washing them down, so you cannot legally fog or mist or smoke food production areas.

Destruction of insects in food areas is therefore limited to the use of physical control methods or fumigant gases. For total pest control, it is not enough to target only visible insect infestations, it is the real unseen juvenile stages of infestations that represent the greatest risk to food production and therefore most need to be addressed.

Physical means of insect control are very varied and include physical cleaning, electric fly killers, humidity control, physical smashing of insects as in flour mill entoleters, removal of insects by sieving and sifting, and of extremes of temperature such as cold and heat. It has been said that some Canadian mills used to simply open up the factory to the −40 °C Canadian winter to kill off insects as a means of control, but extremes of cold and humidity together can cause as much of a problem to electronic machinery and building infrastructure as extremes of heat.

However, whilst extremes of heat will cause damage, controlled use of heat is in many ways the perfect means of controlling insects inside machinery and areas where normal insecticides and cleaning cannot reach. There are three ways of reaching inside inaccessible areas like the inside of food production machinery, stacked commodities, commodities in bins, packed transport containers, timber or furniture to kill insects, and these are:

1. Gases, whereby toxic molecules actively penetrate through most non-metallic materials by diffusion. Examples of this have been historically hydrogen cyanide, ethylene oxide, ethylene dichloride, ethylene dibromide, carbon tetrachloride, carbon dioxide, nitrogen, phosphine, sulfuryl fluoride and methyl bromide.
2. Radiation and microwaves, whereby high energy particles (normally gamma particles, X-rays or microwaves) penetrate any material and when a collision is made they give off

their energy to whatever they collide with, causing damage or heating. However, this is impracticable in most circumstances.

3. Extremes of temperature, either of heat or cold. Extremes of heat are achieved by increased energy being applied to the infested area by convection, conduction or radiation or any combination of the above to bring it up to a temperature lethal to insect life and the destruction of protein structure. Extremes of cold are achieved by the application of cold by convection or conduction and by allowing residual heat to radiate out of the product until insects have reached a stage where life processes stop and cell membranes are split by the expansion of water in the cells caused by the formation of ice. As many cells may be high in solute concentration, the freezing point may be far below zero degrees Celsius. Freezing may also be resisted by slow continued exothermic metabolic processes at a cellular level that continue to give out energy to resist freezing.

Finally we come to the use of chemical pesticides covered earlier, but as can be seen above, there are many things that can be done before you have to resort to the use of chemicals. Not using chemical pesticides is of course essential in the production of organic food. Some pesticides however, when properly targeted and responsibly used rightly, have a role to play and may be used as a last resort to tackle either infestations that have got out of control or for emergency insect control. They will normally form a significant part of an integrated pest management programme.

Most pest control management is reliant on the use of fumigants to control insects inside bulk commodities, not because there are no alternatives (these are discussed in a separate chapter), but because there has not been the investment in the infrastructure and organisational processes required to make them work in industry. There are indeed many ways in which we can use heat to control insects inside many commodities, and CO_2 or N_2 in sealed bins are viable alternatives to methyl bromide or phosphine. However fumigation takes longer, and there are not currently anywhere near enough fumigation facilities around to deal with the demand.

A fumigant such as sulfuryl fluoride is a highly penetrating molecule and is gaseous at the right temperature. However, its egg killing (oocidal) properties are relatively poor compared to methyl bromide or phosphine. Very high concentrations are needed in comparison to a methyl bromide treatment, with up to five times the amount of fumigant having to be used in some cases. Coupled with this, its mammalian toxicity compared to methyl bromide is very high, making it potentially very dangerous to use, both for operators and bystanders. The high concentrations required inside the building to kill insect eggs lead to a steep diffusion gradient which in turn can lead to high rates of diffusion leakage through the structure of buildings and through the slightest of cracks and deficiencies in sealing. The manufacturers recognise this and demand high standards of their licensed fumigators. Super sealing buildings is part of the fumigation process, but managing the safe venting of several tonnes of highly toxic gas in an urban environment under still, heavy weather conditions is potentially hazardous. By comparison, heat treatment can be seen as a much safer and manageable option. It does not require super sealing or a great deal of preparation. Another significant advantage is that at the end of the treatment no venting period is required, meaning that people can enter the building at any time.

References and further reading

Bell, C. (2014), 'Pest control of stored food products: insects and mites', in Lelieveld, H., Holah, J. and Napper, D. (Eds.), *Hygiene in food processing: principles and practice* (Second edition), Woodhead Publishing Limited, Cambridge, UK.

Heaps, J. (Ed.) (2006), *Insect management for food storage and processing* (Second edition), AACC Publishers, USA.

Hill, D. (2002), *Pests of stored foodstuffs and their control*, Springer Publishers, The Netherlands.

Kloosterman, L. and Mager, K. (2014), 'Pest control in food businesses: an introduction', in Lelieveld, H., Holah, J. and Napper, D. (Eds.), *Hygiene in food processing: principles and practice* (Second edition), Woodhead Publishing Limited, Cambridge, UK.

Morland, D. and Hedoes, S. (Eds.) (2004), *The Mallis handbook of pest control* (Ninth edition), GIE Media Inc., USA.

Subramanvam, B. (Ed.) (1995), *Integrated management of insects in stored products*, CRC Press, Boca Raton, USA.

Part Two

Applications in food

Heat treatment of insect pests in food production machinery

4

4.1 Introduction

Food production machinery is perhaps the most vulnerable environment in terms of long-term chronic insect infestation. A large variety of stored product insects (SPI) can become a problem in food machinery, such as the flour beetle (*Tribolium* spp), biscuit beetle (*Stegobium paniceum*), merchant grain beetle (*Oraeziphilus mercator*), mill moth (*Ephestia kuehniella*) and the Indian meal moth (*Plodia interpunctella*). This chapter also covers cigarette production machinery infested with tobacco beetle (*Lasioderma serricorne*).

Spot heat treatment of specific machinery has many advantages over space treatments of whole buildings (see Figure 4.1 and Figure 4.2). It is much cheaper and less disruptive; indeed, if machinery not integral to the main production processes can be shut down and isolated, work can be carried out during normal working hours. In such cases it may be necessary to explain to workers what is going on, in order to address any concerns about insect control going on in working hours and environments.

4.2 Treating different types of food production machinery

For the purposes of heat treatment, food production machinery can be generally divided into two types: 'boxed' and 'open'.

Boxed machinery includes items such as ovens, provers and coolers, industrial dishwashers, mills, elevators and chain link (but not screw) conveyors. These are usually reasonably straightforward to treat with ducted air systems, with air input at one end or at the base and vents at the opposite end or top. In most cases the outside of the machinery will still need to be insulated (e.g. using plastic sheeting), but this can be done at leisure as the inside is being heated up. Likewise, temperature sensors may be placed in position as the machinery is being heated up.

Open machinery such as band conveyors, packers and mill purifiers, etc., need to be insulated prior to treatment. Screw conveyors should also be included in this category, because it is not possible to pass hot air through a closed screw conveyor for any significant length of time due to air resistance by the screw. However,

Heat Treatment for Insect Control. http://dx.doi.org/10.1533/9780857097811.2.31

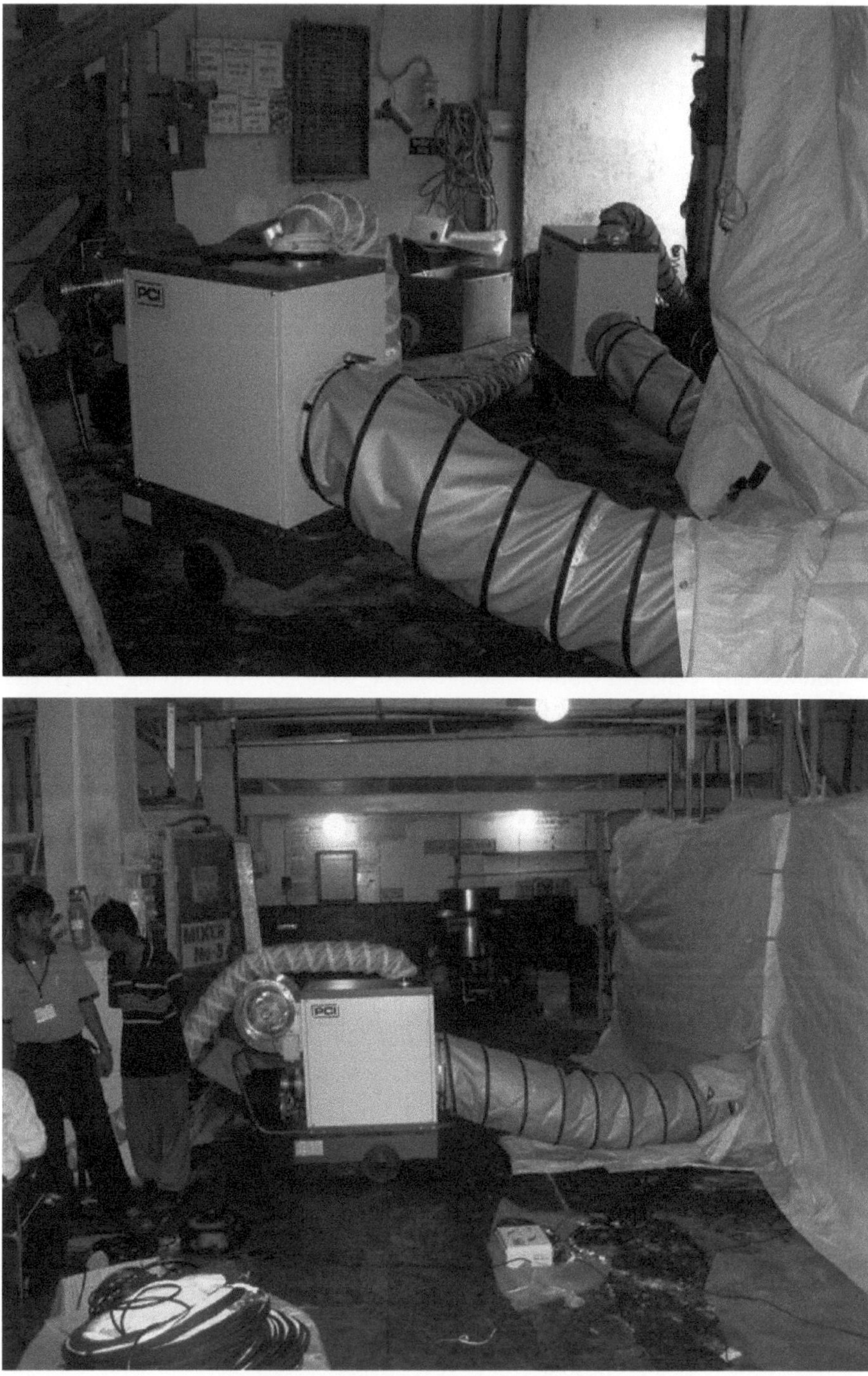

Figure 4.1 Heat treatment of machinery.

Figure 4.2 Heat treatment of storage facilities.

before insulating the machinery, the air distribution ducting must be put in position. This usually includes a plenum box to which shorter flexible ducts are attached, allowing easy adjustments to be made during the course of the treatment as required. As it is not easy or practical to get under the sheets to position the temperature sensors after insulating the machinery, these too are usually located prior to sheeting (see Figure 4.3). It is not recommended to attempt to insert sensors once hot air input has started.

For long production lines and conveyors it may be necessary to do the job in stages, starting heating at one end and, as one section has completed its time at the required temperature, moving the input ducting along to treat the next section, which will have been pre-heated by the waste heat from preceding sections in the line. This is explained in more detail in the section on conveyors and elevators below.

Long, thin structures in either the horizontal or vertical plain such as worm conveyors or elevators should be insulated with polythene or bubble wrap rolled out along their length (or height) and rolled and clipped at intervals to form a poly tunnel covering the whole piece of equipment (see Figure 4.3). The hot air is then passed along the outside (and inside as well if possible, e.g. in the case of elevators) to heat up the metal. As for long production lines, the air input locations may be progressed along/up the conveyor during the course of the heat treatment, if necessary.

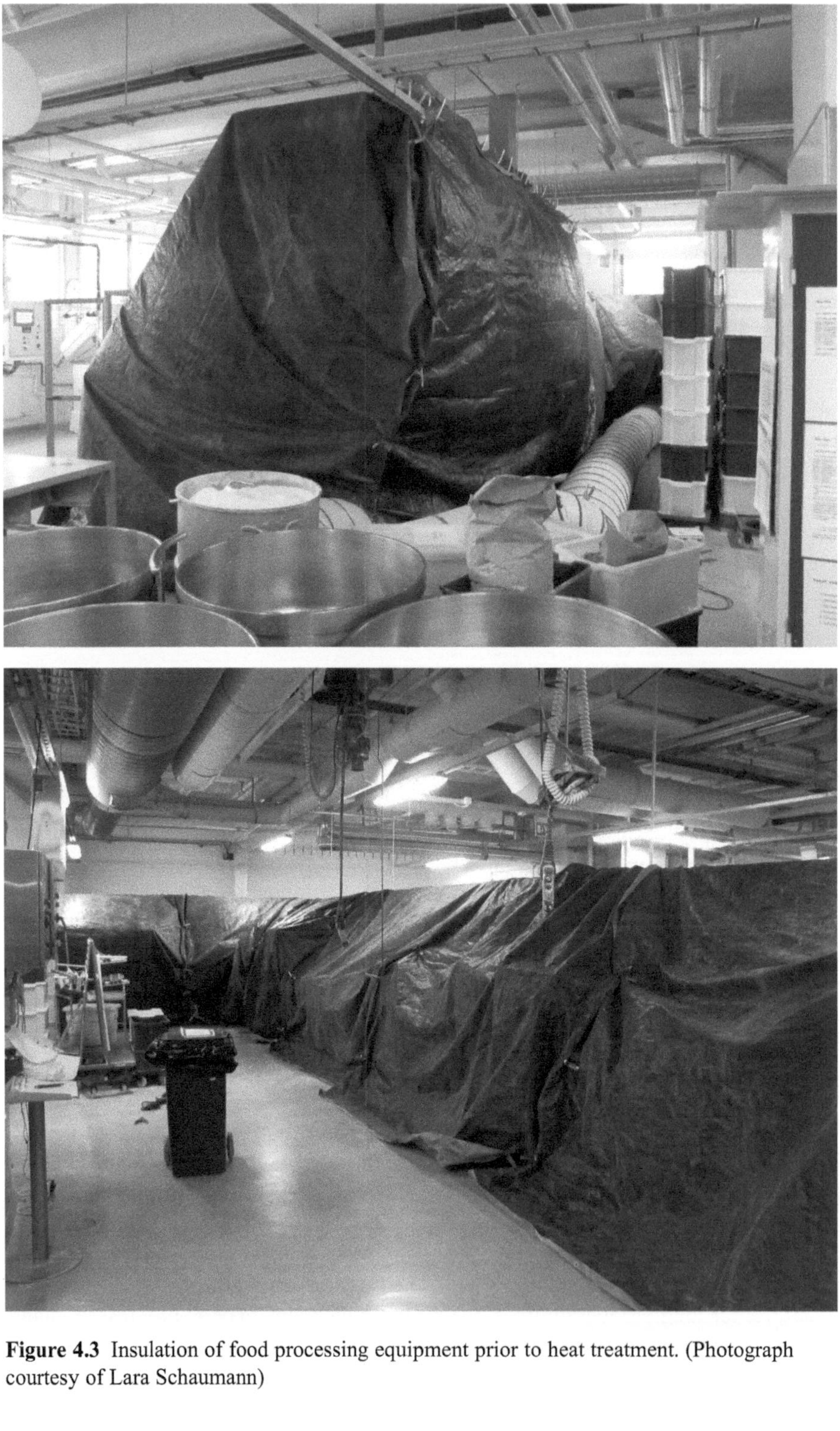

Figure 4.3 Insulation of food processing equipment prior to heat treatment. (Photograph courtesy of Lara Schaumann)

4.3 Dealing with heat-sensitive components

Food production machinery is often highly complicated, containing complex components such as electronic sensors, motors, valves and microchips, and factory managers and engineers are often rightly concerned about the potential effects of heat on these expensive pieces of equipment. There is a very high risk that severe heat treatment (i.e. beyond site-specific maximum temperature thresholds) may cause damage. Damage to machinery can be avoided by the simple expedient of discussing treatments in advance with engineers and taking care to remove or protect any sensitive machinery that could be at risk.

In fact most electronic components are physically quite heat resistant given the range of applications they are typically used for. Most aircraft are designed for safe operating temperatures of up to 65 °C as they have to operate in desert areas with plenty of safety margin to spare (though this should be checked on a case-by-case basis for specific aircraft heat treatment jobs). However, as electrical resistance increases with temperature, equipment may not function properly at high temperatures, or may overheat. The golden rule for heat treating machinery is to turn off, lock off and isolate all electrical power, for health and safety reasons.

Some lightweight plastics and synthetic materials may soften slightly during heat treatments, particularly low-pressure air lines controlling valves, etc. All compressed-air-to-air lines must therefore be turned off prior to treatment, and input air must never be directed onto plastic surfaces. Insulating material may be used to protect vulnerable materials from the direct input of air, which is the hottest part of the operation. If any areas are of concern to clients, they can be protected during the course of a heat treatment by covering with a damp cloth and inserting a temperature sensor with an agreed maximum temperature specified by a suitably qualified engineer.

Sometimes the heat treatment technician will be presented with a completely unknown product, for example where glue or resins are involved in the construction of machinery. An example of this is the case of flour mill plan sifter sieve trays, which use nylon meshes secured by a variety of different types of glue, some of which are heat resistant up to 60 °C and others which are not. In such cases, the surveyor should identify the initial risk and take a test sample for exposure to the target temperatures. This can be done very simply in a household electric oven, which makes an ideal test bed.

4.4 Treating conveyors and elevator systems

Bins and silos are loaded and discharged either by fast-flowing pneumatic air systems, which are not often infested with insects due to the lack of suitable undisturbed areas for breeding (although it is not unheard of for insects to find homes in cracks and rough joins between elements of piping or even in rotary seals). However, old-fashioned bucket elevators, screw/worm and chain drag/redler conveyors often become significantly infested with pests such as stored product beetles, moths, mites and booklice. The degree of exhaust (dust extraction) contributes significantly to the level and type of infestation present. However, conveyors and elevators are quite straightforward to heat treat.

4.4.1 Elevators

The procedure for heat treatment of elevators is as follows:

1. Survey and carry out risk assessment, ensure elevator is locked off (isolated), and cleaned as thoroughly as possible and check that no stock is left in the bottom. Ensure that all slides from the elevator at the top are open to allow air out and open any inspection hatches both at the top and at intermediate stages.
2. Remove as many slides as possible at the base of the elevator and clean/spray with residual insecticide, or clean and leave within the heat treatment area.
3. Install temperature sensors inside the elevator, in the vertical sections if possible and at the top (it is especially important that there are sensors placed at the top of the elevator).
4. Install the input ducting, ideally via four small ducts: one directed up each vertical column and one up the outside of each section.
5. Cover the elevator with polythene sheeting on a roll. Two-metre wide polythene wrapped around the elevator and rolled and clipped at intervals of roughly one metre will be sufficient in most cases. The idea is to create a long, vertical polytunnel, so that hot air passes both up the outside of the elevator and inside it as freely as possible.
6. Begin the heating process. The bottom sections will heat up more quickly. Once a section has reached 50 °C for a minimum of one hour and there are no live insects to be seen, the input ducting can be moved from the bottom progressively up the elevator. If there are no additional inspection hatches up the elevator to input air, leave the bottom internal ducts in place and just extend the external input ducts progressively up the elevator until all sensors have reached 50 °C for one hour and, most importantly, the internal air out sensor has read 50 °C for two hours, or for one hour if it is genuinely reading the temperature of the metal (or wood) of the elevator top/discharge spout.
7. As always, do a visual check for live and dead insects before switching off the heat.

4.4.2 Conveyors (excluding blow-lines)

As with elevators, conveyors are not difficult to treat. However, some types are more difficult than others. In screw conveyors the corkscrew effect on the air flow creates air resistance, which combined with a small cross-section area means that almost no air will make its way down the conveyor inside. This means that the heater is likely to overheat and cut out. Unlike bucket elevators, there is no natural vertical air convection (rising hot air) to assist the heat treatment technician either.

For worm conveyors the procedure is as follows:

1. Survey and risk assessment – ensure conveyors are treated as part of an integrated heat treatment, including associated machinery. If treatment is to be carried out over several days or in parts, ensure that the order of the treatments is in the direction of stock flow through the system. Ensure that the system is cleaned and electrically isolated.
2. Remove as many lids as possible, at least every two metres.
3. Install internal temperature sensors every five metres or so in any enclosed sections 'downwind' from the air flow. Sensors should be attached to the metal (or wood) of the conveyor.

In most conveyors temperature sensors may be wedged between the worms and the wall of the conveyor so that they measure the temperature of the metal without the need for tape and the consequent risk of product contamination.

4. Install a single broad input duct at the start of the conveyor system and sheet-over the conveyor, unrolling a two-metre roll of lightweight polythene or bubble wrap and rolling and clipping along the length to form a polytunnel.
5. Switch on the heat. Once the first sections have heated up to 50 °C for one hour in the coldest part, check for any surviving insects before moving the input ducting down the line, and so on until complete.

The procedure for treating chain drag/redler conveyors is as follows:

1. Survey and risk assessment – as for worm conveyors.
2. Remove the end plate of the conveyor to expose the twin levels of the conveyor. Engineers will be required to do this. Unlike worm conveyors, air will travel quite considerable lengths down the inside of a redler conveyor if it is large enough in cross-section.
3. If the length of conveyor is short (<4 m), the cross-section area large enough (>0.25 m^2) and the end plates removable, then it may be sufficient just to blow air down the inside of the conveyor and heat up to 52 °C for one hour without sheeting the outside (this can be sprayed to kill any surface-living or erupting insects). If the conveyor is long, narrow or for additional effectiveness, create a polytunnel as for screw conveyors and heat up.
4. For a long redler conveyor of small cross-section area, create a polytunnel as above and heat up entirely from the outside, aiming for 52 °C for one hour external (non air-influenced) temperature readings or 50 °C if internal measurements are possible.
5. Switch on the heat. Once the first sections have heated up to 52 °C for one hour in the coldest part, check for any surviving insects before moving the input ducting down the line, and so on until complete.

Blowlines and pipes can be treated in the same way as a small redler conveyor.

4.5 Monitoring temperature levels and distribution during heat treatment

It is crucial to the success of any heat treatment to achieve and maintain an even distribution of lethal temperatures in the areas where the insects are living or may try to use as refuges, and the correct measuring of temperature is vital. The operator in charge must be able to understand the movement of air within the stack and the rates at which different areas heat up. During the survey he will plan a route for the hot air to be ducted through the stack and may decide whether to use only one treatment or divide the job into stages, heat treating different sections in sequence.

However, without information on core temperatures the operator will not be able to monitor or control the movement of hot air within the stack. As many fixed temperature sensors as possible should be used to monitor the progress of the job, positioned away from the air flow in potential cold spots (see Figure 4.4). Cold spots often occur low down in the area to be treated, but can also appear in pockets of still air under the covering sheet. Fixed temperature sensors must be placed in a grid pattern to cover all parameters of the stack, attached to the heaviest pieces of metal (which

Figure 4.4 Heat sensor to monitor heat treatment of food machinery.

will take longest to heat up), inside hollow voids and cabinets (open doors where possible) and if cleaning is not possible, inside the base of the thickest food residues that can be found. Fixed-temperature sensors are crucial for monitoring the rates of increase in temperature to determine whether the air movement and heat distribution is uniform. A single heat treatment job typically uses between eight and 16 sensors.

Temperature logging on a real-time graph is crucial in heat treatments as it allows the heat treatment technician to see how the job is progressing and tells him (by observing slow rates of increase) where additional heat may be required. In addition, monitoring areas where the temperature is increasing rapidly can indicate potential for redistribution of the heat input to ensure uniform heating.

Mobile infrared spot temperature readers and heat cameras are a useful tool for supporting the operator, but should never be used as a main form of monitoring as they do not provide a heat/time record or record rates of temperature increase. However, they can be very useful in identifying potential cold spots.

Entering the stack to check for cold spots or cold air is important during the treatment, as well as a final manual check for live insects prior to turning off the heat at the end of a treatment. Care should be taken to maintain health and safety when entering a heat treatment zone, as heat exhaustion can be fatal. It is recommended to wear gloves and other protective clothing if it is necessary to be in contact with equipment (e.g. ladders), drink plenty of fluids and leave the area immediately if feeling unwell. Technicians should also work in pairs in order to avoid missing the symptoms of heat exhaustion. Metal rings and necklaces should be removed or moved outside clothing if a person enters a heat treatment area as they will heat up very rapidly and may cause localised burning. It should be noted that such items would not normally be permitted to be worn in a food production area anyway.

4.6 Additional measures to ensure the effectiveness of heat treatment

In all cases of infested machinery to be heat treated, pre-cleaning to remove all significant food residues is essential. Any remaining residues of concern should be monitored for temperature penetration at the deepest points. Post-treatment cleaning is also essential. As the equipment (especially that made of metal) heats up, insects living inside previously safe and inaccessible harbourages often emerge in huge numbers, often way beyond any levels expected by the client, and die on conveyor belts and food contact surfaces. These need to be removed and the equipment thoroughly cleaned before food production can resume.

The results of heat treatment can be dramatic, especially in the case of food production machinery where there has been a history of insecticidal fogging on site. Toxic air environments (e.g. fogging with pyrethroids) tend to force breeding insect populations deep into the depths of machinery, paradoxically making it more likely that the insects will survive in inaccessible places and contaminate food products. Often the full extent of an infestation only becomes apparent once the heat treatment has been carried out. As heat passes through the food production machinery, the fabric of the environment in which the insects are living becomes increasingly intolerable and insects can emerge in high numbers. Fast-moving cockroaches quickly become agitated by heat and do their best to exit the heat treatment area when temperatures start to rise above 40 °C.

It is thus most important to make provision for containing the erupting infestation of insects. Where permitted, it is always advantageous to spray insecticide around and under food production machinery prior to heat treatment to kill adult insects trying to escape the heat. If insecticides are not permitted, spraying with vegetable oil or laying down sticky paper around machine footings to trap insects can prevent re-infestation of machinery by insects on any sections of floor that do not reach the target temperature. The undersides of sheets and heat resistant tarpaulins can also be sprayed prior to treatment so that cockroaches and other insects cannot find harbourage in the cool folds of the sheets. Two case studies illustrate these points.

4.6.1 *Case 1: an example of the importance of cleaning*

The scale of an insect infestation is not always evident and may require extra time and resources to address. Heat treatment of 34 pieces of machinery in a bakery using the Thermokil heat treatment system took place over two evenings. The bakery manager was aware that there was a persistent and widespread 'small' infestation of flour beetle and mill moth. The insects were not emerging in high numbers, but insect tracks and little pockets of moth webbing were evident. Heat treatment was arranged for the Friday and Saturday nights, after production had finished for the day but before 04:00 when the next shift was scheduled to begin. The heat treatment was begun on the machinery and the bakery manager went off to his bed in preparation for his 04:00 start the next day. At 22:00 the leader of the pest control team had reluctantly to wake the manager to inform him that a cleaning team was needed prior to starting production, as the machines were covered with thousands of dead flour beetles and biscuit beetles that had emerged during the heat treatment process.

4.6.2 Case 2: an example of floor spraying taking effect

In Denmark, a heat treatment training programme for pest control technicians was arranged in a large bakery. The purpose of the training programme was to show technicians how to sheet up machinery and install temperature sensors, and to demonstrate the concept of pre-spraying around machine footings prior to treatment. The course was divided into two groups, with one team selecting a long bread cooler to sheet-up and start heating and the second team a large automated mixing complex with a known mixed German and Oriental cockroach problem, complete with a caged off area housing automated machinery for the tipping and mixing of ingredients. The area had been repeatedly sprayed and there were numerous residues of insecticide gel, some recent, where the regular pest control technician had attempted to deal with the insects before regular wet washing removed any residual insecticide. Within an hour, the team had the floor area sprayed and the treatment area roughly sheeted over. The heat treatment equipment was switched on and the sealing refined. With the afternoon training session drawing to a close, there was no time to complete the actual treatment, but temperatures in the mixer complex were averaging 40–45 °C when some of the team decided to go inside the stack to see what was going on. They were greeted by the site of hundreds of cockroaches running around trying to escape the hot machinery and bailing out onto the toxic floor where they were killed by the insecticide and heat combined (see Figure 4.5).

4.7 Risk assessment to ensure the effectiveness of heat treatment

It is most important when treating machinery to take into account the legitimate fears of engineers and production managers about the risks of exposing often very expensive machinery and electronics to increased temperatures. There are a few golden rules to observe before and during heat treatment, which should be addressed during the pre-treatment survey:

- Always ask the production and/or engineering manager what pieces of equipment he is most concerned about. Any fixed items of concern can be either kept outside the sheeted-off area (e.g. flat screens on stems, electrical control boxes) or can be covered and kept cool during the heat treatment with a damp cloth. Removable items such as inkjet printer cartridges can of course be taken away while treatment is in progress. In any case, where there is concern, the heat treatment operator in charge should agree a safe maximum temperature with the production manager or engineer and ensure that this is not exceeded during the course of the treatment. He can also place temperature sensors on any at-risk items, agreed with and witnessed by the management. A maximum target temperature of 60 °C is normally sufficient, although some glues and light plastics may soften even at this temperature. If in doubt it is good practice to test some items prior to treatment in an oven, or simply keep them cool. In any case one should never exceed 70 °C when treating machinery, as damage is likely at this temperature.
- Fire detection and suppression systems are often installed in and around food production machinery, ranging from simple heat detectors linked to sprinklers to full chemical fire suppression systems. Accidentally setting one of these off can incur costs of many thousands of pounds. Before carrying out any heat treatment, the operator in charge must enquire whether any such systems are in place and take measures to prevent activation. Electronic detection systems can be isolated and sprinkler heads in risk areas covered with a damp tissue inside an inverted plastic cup with a temperature sensor placed inside. Sprinkler heads

Figure 4.5 Dead cockroaches after combined heat/insecticidal treatment.

can be activated as low as 54 °C although this can vary up to 75 °C depending on type. Such variables must always be checked during the heat treatment survey.

- Many types of electrical equipment are quite heat durable *when switched off*, however they will fail if working at high temperatures. A maximum working temperature will usually be specified by the manufacturer. These temperatures are not necessarily indicative of the real operating parameters but are the minimum the manufacturer is willing to sign up to for safety reasons. Most electronic equipment can withstand reasonably high temperatures, however it is important to switch equipment off and consult the site manager about what s/he might wish to be removed prior to starting a job.
- Some materials do soften during heat treatment, particularly some glues, adhesives and plastics, although they harden up again as they cool. Therefore if there is a risk the following precautions may be taken:
  (a) Always turn off small pneumatic control lines. Even at quite low temperatures, air lines may be blown off if kept under pressure, so turn off all compressed air supplies.
  (b) Never direct heat onto plastics, particularly lighter plastics. Ambient temperatures up to 65 °C are normally safe but with some direct input temperatures potentially as high as 70–80 °C it is always good practice to make sure any hot input air is directed onto a safe first surface (e.g. metal) or deflected using cardboard box sections into a safe area and away from any risk items.
- Machinery must be locked-off and isolated, and the key to the isolators kept with the operator in charge. The consequences of machinery starting up automatically while operators are inside could be fatal.
- If temperature sensors are secured using tape, there is a risk of product contamination if tape is left inside the machinery. The golden rule is always to use one piece of tape only, count one in and count one out. If possible use metallic tape, which is easily visible and will show up on a metal detector.
- One of the biggest risks when using diesel heaters is the presence of fuel oil in a production area, and all precautions must be taken to prevent spillage. All fuel must be kept in fully enclosed containers, which must then be placed in a bund to contain any leakage. In the base of the bund there should be an absorbent material. The bund should then be placed on a plastic cover surrounded by a fuel spill boom. A separate fuel spill kit and fire extinguisher should be on hand. Modern diesel heaters such as the Kroll units used for most heat treatments have all the safety features you would find on an oil-fired central heating burner, and we are all happy to go to sleep with one of these heating our houses at night.
- Electrical safety must always be considered, particularly where there is a risk of sprinklers being activated. All 220-volt equipment must always be used with an RCD protection device. Greater precautions must be taken when using 18 kW, 440 v heating devices.
- Sheeting-over machinery and silos often involves working at heights while handling sheets up to 25 kg in weight. It is not safe to be pulling up a 25 kg sheet with one hand while clinging precariously to the top of a ladder with the other. The surveyor of the job must work out in advance how he envisages sheeting-over the machinery and ensure the provision of safe access equipment ahead of the job. Some heavy heat-resistant tarpaulins have eyelets along the edge, allowing them to be pulled into place with para-cord or equivalent.
- When carrying out heat treatments of machinery during normal production hours, it is often likely that air input or output exhaust ducting may obstruct normal pedestrian or fork lift traffic routes. These should be identified during the survey and provisions made to accommodate production requirements.

References and further reading

Bell, C. (2014), 'Pest control of stored food products: insects and mites', in Lelieveld, H., Holah, J. and Napper, D. (Eds.), *Hygiene in food processing: principles and practice* (Second edition), Woodhead Publishing Limited, Cambridge, UK.

Heaps, J. (Ed.) (2006), *Insect management for food storage and processing* (Second edition), AACC Publishers, USA.

Hill, D. (2002), *Pests of stored foodstuffs and their control*, Springer Publishers, The Netherlands.

Kloosterman, L. and Mager, K. (2014), 'Pest control in food businesses: an introduction', in Lelieveld, H., Holah, J. and Napper, D. (Eds.), *Hygiene in food processing: principles and practice* (Second edition), Woodhead Publishing Limited, Cambridge, UK.

Morland, D. and Hedoes, S. (Eds.) (2004), *The Mallis handbook of pest control* (Ninth edition), GIE Media Inc., USA.

Subramanvam, B. (Ed.) (1995), *Integrated management of insects in stored products*, CRC Press, Boca Raton, USA.

Thermokill heat treatment manual.

Heat treatment of insect pests in food storage facilities

5

5.1 Introduction

Bins and silos such as those used to store grain, flour and other foodstuffs must remain free of insect infestation, particularly if they are to be filled with finished product. They can be ready-made fumigation chambers when full of stock. Metal bins are particularly suitable as they can be easily sealed and fumigated with phosphine or CO_2 for long periods of time. However, it is not always possible to take an empty bin out of service for the long periods required for fumigation with phosphine or CO_2, particularly where that bin is an integral part of a production cycle. *Ephestia* spp. moths, particularly mill moth (*Ephestia kuehniella*), produce vast quantities of sticky webbing which can provide a framework for infestation with other insects such as flour beetles of various kinds (e.g. *Tribolium* spp. and *Gnathocerus cornutus*). In these circumstances, heat treatment of the empty bins and their associated feed and discharge conveyors, duct collectors and elevators can be very useful.

5.2 Heat input and sensor positioning

It would seem at first glance that heat treatment of bins should be simple. However, bins and silos come in such a vast array of sizes and designs that a variety of completely different techniques have been developed for use in different situations. The principle is the same as for all heat treatments: hot air is blown over all of the surfaces that need treating to bring all areas of potential infestation up to or above 50 °C.

Like food production machinery, almost all silos need to be sheeted-over unless already insulated, as is the case with outdoor silos (Figure 5.1). Although it may not be possible to fully cover the entire silo exterior, the heat treatment technician should cover as much as possible so that metal sides are not exposed to, and cooled by, external ambient air. Silo tops and bottoms are generally accessible and simple to sheet-over, but sheeting must be loose to allow displaced air to escape as required.

The next limiting factor is finding an access point large enough to apply a rapid flow of hot ducted air (65 °C to 80 °C). However, in some cases a screw worm discharger may be positioned at the bottom of the silo restricting air flow, which would make this point unsuitable for hot air input. In such cases, hot air can be applied from the top and vented out the bottom, or applied via side access panels. In the most simple case of a direct load bulk out-loading silo, there are large enough access points for

Heat Treatment for Insect Control. http://dx.doi.org/10.1533/9780857097811.2.45

Figure 5.1 Sheeting over bin tops.

air input, allowing hot air to be blown up through the bins to the top and vented out both via the open hatch under the sheeted-over top and through the dust collector/air exhaust system (see Figure 5.2).

As with all heat treatments, temperature monitoring is crucial, and in silos there is a standard protocol for the positioning of sensors. In all silos there should be around six temperature sensors in the following locations:

T1 – on top of silo, but under sheet;
T2 – inside dust collector (if applicable);
T3 – air temp at top of silo (air in if applied from above);
T4 – inside base of silo (lying on the bottom of the silo or inside the discharger);
T5 – on outside of base of silo under sheets (opposite side to discharger exit);
T6 – air temp (air out if air applied from above).

T4 is very important in flour silos in particular. As hot air dries out the silo, flour sticking to the sides loses its adhesive property and may drop to the bottom, blocking the air outlet point and providing a nice cool refuge for insects. If this happens, the sensor in the bottom will quickly show a reduction in temperature and the operator can take measures to clear this accumulated flour. Heat treatment could potentially be applied to routinely clear the bulk of accumulated damp and/or mouldy flour from the outside of bins, instead of having to scrape it off manually.

Additional sensors may be applied to measure the temperatures on the outsides of bins (under the sheets). This provides an efficient means of determining that the internal temperature is high enough when hot air is applied inside the bin, as the outside of the bin will be colder than the inside. Temperatures should always be measured in the coldest spot.

As with most situations, the better the preparation, the better the results. Thorough cleaning prior to heat treatment of bins and silos will result in a quicker and more effective job and will help to prevent the potential problem of damp stock falling off the walls and blocking air movement at the bottom of the bin.

5.3 Heat treatment protocols for different types of bin/silo

As there are a huge variety of sizes and designs of bins and silos, we have suggested a variety of techniques to allow successful heat treatment of most of the more common designs. If there is no large enough access point to introduce hot air into the silo, an engineering input may well be required; for example to remove a section of the hopper bottom. The heat treatment technician should not be afraid to stipulate this input from the client.

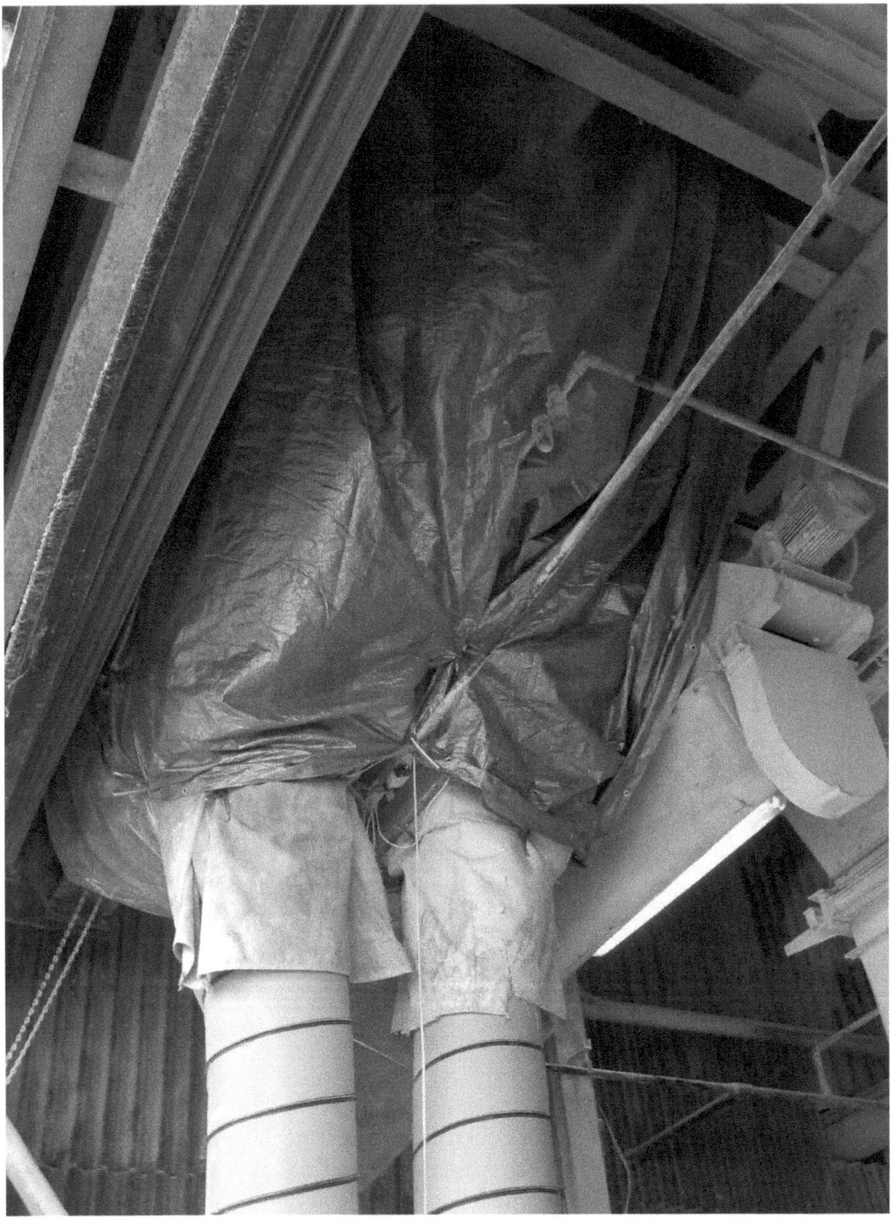

Figure 5.2 Access point for heat treatment of a silo.

5.3.1 Bin protocol 1: large bulk outloading silo

This type of silo can be unloaded either via direct drop into trucks or via a removal slide mechanism giving at least 0.25 m^2 of unrestricted access for air input at the bottom and with an open lid for air exhaust at the top. Key steps are:

1. Carry out preliminary survey and risk assessment.
2. Clean the silo (if possible).
3. Open the top lid.
4. Attach or insert air input duct at the bottom.
5. Position six temperature sensors as described above.
6. Switch on the heat.
7. Sheet-over the top of the bin first (allowing space for air to escape under loose fitting sheets).
8. Sheet-over the bottom of the bin.
9. Sheet the sides (as far as possible).
10. Monitor rates of temperature increase, paying particular attention to the sensors in the bottom of the bin on the outside of the hopper (under sheets). When all the sensors have reached 52 °C and maintained that temperature for one hour, take samples from inside the bottom of the bin and inspect bin top for signs of life before switching off the heat.

5.3.2 Bin protocol 2: large silo with worm discharger at the bottom and a side access panel for internal access

Key steps are:

1. Carry out preliminary survey and risk assessment.
2. Clean the silo (if possible).
3. Open the top lid.
4. Attach or insert an air input duct via the side panel, directing air downwards to the bottom of the silo.
5. Insert or attach a vacuum cleaner hose into the bottom of the bin below the air duct.
6. Block air from moving out of the bin via the dischargers by stuffing lightweight polythene into the discharger line where access permits, leaving the bin lid open.
7. Position six temperature sensors as described above.
8. Switch on the heat.
9. Sheet-over the top of the bin first, covering the open bin lid.
10. Sheet-over the bottom of the bin.
11. Sheet the sides as far as possible.
12. Monitor rates of temperature increase, paying particular attention to the sensors in the bottom of bin on the outside of the hopper (under the sheets). When the top three sensors all have reached 50 °C and maintained this temperature for one hour, inspect the bin top for signs of insect life. Ensure that there are no signs of life before moving on to the next step.
13. Remove polythene stuffing from the discharger conveyors, allowing air to exhaust at the bottom and paying attention to changes in temperature shown by the sensors inside the discharger, at the bottom of the bin and on the outside of the bin hopper.
14. Close bin lid, stopping air from exiting at the top of the bin. The air that was originally moving upwards through the bin now has no other way to go but down and out through the

dischargers. Any insects that survived the heating of the top of the silo by dropping to the bottom should now be subjected to the full force of the heaters and killed at the bottom. Rising heat inside will ensure the temperature at the top of the bin does not drop, but will remain at around 50 °C for an extra hour.

15. If temperature sensors at the bottom do not show a sharp increase, it could be because stock from the sides of the bins has dropped onto them and blocked the air outlet route. Use the vacuum cleaner to suck out any such loose stock and clear the route for air to escape from the bottom.
16. Temperatures at the top will remain close to the maximum temperature for the job due to natural convection patterns within the bin as hotter air rises and displaces cooling air at the top.
17. When all sensors have maintained 50 °C for one hour in the coldest spot, take samples from the base of the bin and from the discharger and check the ledges on the outside of the bin, the bin top and inside any dust collectors for signs of life. Ensure that there are no signs of life before switching the heat off.
18. It is possible that parts of the discharge system may not be large enough to allow sufficient air flow through them to heat up quickly. In these cases, it is best practice to insulate these areas using sheeting and heat them from outside using a small supplementary heater at the bottom. A small electric heater may be sufficient for this in many cases. When applying heat to the outside of a conveyor system, the temperature should be measured on the inside.

5.3.3 Bin protocol 3: small to medium sized bin

This type of bin has no access at the bottom or sides for cleaning and uses a rotary seal discharger feeding into a pneumatic discharge system. Access is only via a lid at the top of the bin. The bin is fed by an overhead conveyor system. Key steps are:

1. Carry out a preliminary survey and risk assessment.
2. Clean the bin if possible.
3. Open the top lid and ensure all input bin slides are open to allow air to exit upwards through bin loading conveyors.
4. Position six temperature sensors. Five of these should be placed according to the normal bin profile. However, the sensor that would normally be in the bottom discharge conveyor (effectively the ‘air out’ sensor) should now be placed in the overhead input conveyor, as in this case the air is exiting the system via the input conveyor and to some extent back out of the lid alongside the input duct.
5. Insert the air input duct through the top lid as far down as possible, so that it is directed onto the top of the rotary seal.
6. Switch on the heat.
7. Sheet-over the top of the bin first.
8. Sheet-over the bottom of the bin.
9. Sheet-over the sides as far as possible, and also the top of the input conveyor.
10. When all the sensors have shown a temperature of 52 °C for one hour in the coldest spot, inspect for signs of insect life. Ensure that no signs of life are present before switching off the heat.
11. Because these smaller bins and silos are often part of a conveyor system, equipment both up and down the production line should be inspected and associated conveyors, elevators and dust collectors should be treated as appropriate to prevent re-infestation.

5.3.4 Bin protocol 4: small packing bin in flour mill

This type of bin is open at the top, dropping directly downwards into a pneumatic slide operated packer. The bin in this example is situated in an ATEX dust explosion risk area. Key steps are:

1. Carry out preliminary survey and risk assessment, including identification of the area designated as an ATEX dust explosion risk.
2. Clean the bin and ensure that the surrounding area in the room is clean. In this case, heat treatment should only be carried out during factory downtime to avoid any risk of machinery blowout.
3. Insert input ducting into the top hatch and open the slide at the bottom of the bin around the packer. Situate the heating unit and fan outside the ATEX area.
4. Attach an earthing cable from the bare metal supporting coil within the ducting to a safe earthing point.
5. Clean around the packer and floor area and spray with residual insecticide, or fix a catch bag loosely over the exit spout to catch insects as they are blown downwards through the bin.
6. Attach six temperature sensors according to the normal profile for bins and silos.
7. Switch on the heat.
8. Sheet-over the rest of the outside of the bin.
9. When all sensors have reached and maintained a minimum of 52 °C in the coldest spot for one hour, inspect for signs of insect life. Ensure that no living insects are present before switching the heat off.
10. As these small bins are often part of a bigger production run, detailed inspections may reveal insect life in associated machinery and in the complicated components of the packer – these should be treated either separately or at the same time as the inside of the packing bin.

5.3.5 Bin protocol 5: small bin in bakery

This type of bin may be of fabric construction with no significant access points to apply hot air. There is usually a pneumatic discharge and filling system and a small exhaust duct leading to a dust collection system. Key steps are:

1. Carry out a preliminary survey and risk assessment.
2. Ensure bin is completely empty and agitate the sides of the bin as much as possible during the emptying process to clear any flour or stock adhering to the sides of the bin walls.
3. Attach a set of temperature sensors inside if possible, particularly in the bottom of the bin, but attach the rest to the outside of the bin and to the top at points furthest away from any direct air flow so that they are measuring the temperature of the metal and are not influenced by the hot air. This can be done by taping bubble wrap or cardboard over the sensor after it has been fixed to the metal of the bin with aluminium tape.
4. Set up input ducting at the bottom of the bin in a profile that will allow air to be directed upwards under the sheets to the top. In this example, the bin is being heated from the outside, with the air outlet at the top of the fully sheeted bin.
5. Loosely fit the top sheet to allow air to vent out at the top.
6. When all sensors have reached and maintained a minimum temperature of 52 °C for two hours or 55 °C for one hour, the heat can be switched off. The longer time is to allow for heat to penetrate any accumulations of flour in the bottom of the bin that cannot be detected.

5.4 Troubleshooting

Other awkward situations can occur when treating bins and silos, as there seem to be an almost infinite number of bin designs and locations. Some examples and advice are given below.

It is difficult to treat large silos with restricted access, especially if the silo is situated outside and/or the external temperature is low. For example, treatment of a large outdoor grain silo situated among many others with no close access for a heater should not be attempted in winter, as the job will fail. The technician will need to explain the restrictions to the client and return when the weather is warmer.

If there are no decent sized points of entry for input ducting (bearing in mind that the diameter of the input ducting will be around 450 mm to 550 mm) then the client will need to be prepared to remove either the hopper or the discharge system for heat treatment to take place. The alternatives are to fumigate with phosphine or internal cleaning (which again will require access).

If the heater cannot be located close to the bin entry point, for example because the only point of entry is 30 m up in the air, then the only option is to attempt to push hot air over long distances through insulated flexible ducting as fast as possible. In some cases, particularly where a number of bins require heat treatment over a long period of time, it can often be worthwhile to raise the heater to the required height and leave it there over a period of months, treating the bins in succession as they become available.

The input temperature from the heater needs to be as hot as possible without risking damage to equipment (ducting needs to be suitably heat resistant at the hot end) and the outside temperature needs to be as warm as possible at the anticipated end of the treatment cycle. The technician will need to give consideration to minimising air resistance in the ducting, possibly by using ducting with a slightly larger diameter than the air outlet diameter of the heater. The size of the heater fan is another important consideration. Radial fans are better than axial fans. It may be possible to incorporate an additional booster fan into the air input cycle, or alternatively to help the air inflow process by fitting a fan to the air outlet point, creating a slight negative pressure in the bin. However, the air outlet fan should not be more powerful than the air input fan or there is a risk of collapsing elements of the flexible input ducting.

Insulation is crucial and should be placed both on the outsides of the bins and on the input ducting. The insulation should be monitored and if necessary augmented throughout the course of the heat treatment.

In cold conditions, the timing of the treatment can be important. Remember that in many cases 47–48 °C for four hours will be sufficient for an effective treatment, particularly if the previous ambient temperature was in single digits. However, remember to allow plenty of extra time; and if working outdoors where the elements may affect the job be sure to start early and aim to peak at around lunchtime when the day is at its warmest. It is advisable to work with nature and the natural laws of physics to avoid extra costs in terms of time and fuel.

The biggest hindrance nature can provide for fumigators when working outdoors is wind; for heat treatment technicians, it is rain. Water is a powerful coolant and there is nothing quite like heavy rain pouring down on the sheets covering a silo to wreck the day of a heat treatment technician. When planning a treatment outdoors, pay attention to the weather forecast. Small lengths of exposed input ducting can be covered in polythene sheeting over blankets, or insulating bubble wrap on a roll, but not the vast open sides of a grain silo. It is necessary to ensure that the client is aware of the potential limitations on the job and of the potential costs incurred by difficult jobs. For example, it may be necessary to use a crane to get a large heater to the top of a silo block. If there are a large number of silos to be treated, it may be more cost-effective to leave the heater where it is until the entire job is complete. In this case, a monthly fee for storage of the heater on top of the silos would need to be agreed with the client.

References and further reading

Bell, C. (2014), 'Pest control of stored food products: insects and mites', in Lelieveld, H., Holah, J. and Napper, D. (Eds.), *Hygiene in food processing: principles and practice* (Second edition), Woodhead Publishing Limited, Cambridge, UK.

Heaps, J. (Ed.) (2006), *Insect management for food storage and processing* (Second edition), AACC Publishers, USA.

Hill, D. (2002), *Pests of stored foodstuffs and their control*, Springer Publishers, The Netherlands.

Kloosterman, L. and Mager, K. (2014), 'Pest control in food businesses: an introduction', in Lelieveld, H., Holah, J. and Napper, D. (Eds.), *Hygiene in food processing: principles and practice* (Second edition), Woodhead Publishing Limited, Cambridge, UK.

Morland, D. and Hedoes, S. (Eds.) (2004), *The Mallis handbook of pest control* (Ninth edition), GIE Media Inc., USA.

Subramanvam, B. (Ed.) (1995), *Integrated management of insects in stored products*, CRC Press, Boca Raton, USA.

Heat treatment of insect pests in food materials and other fresh produce

6

6.1 Introduction

One of the golden rules of heat treatment training is to make people aware of the limitations of heat treatment. Key potential problems include presence of excess water, lack of possible air circulation, too much mass or too much insulation. Heat treatment of commodities such as food materials can encompass the last four issues and, in many cases, may be considered non-viable. However, there is certainly a potential market for the heat treatment of commodities, for example control of insects and fungal pathogens in organic products.

This chapter will discuss possible options for the treatment of:

- nuts
- cocoa beans
- pulses
- grains and rice
- dried fruit
- fresh fruit and vegetables
- fresh flowers.

There has been quite a lot of work done on sub-lethal heating of fruit and its effect on ripening, but lethal temperatures to insects seem to result in damage in most cases. Lurie (1998) provides a good summary of this work.

One of the starting problems is that each commodity is different and there are no catch-all rules that can be applied to everything. However, the following issues apply to nearly all commodities:

- What is the impact of temperatures lethal to insects on the commodity and can this impact be reduced or eliminated by different temperature × time profiles?
- How is the commodity packed or handled? Is there a stage in the production or handling process where heat treatment may be feasible?
- What is the impact of water loss or gain (depending on treatment type) during heat treatment?
- Are there any other risks associated with heat treatment, e.g. mould growth due to water fall out during cooling, or physical damage to fruit and vegetables?
- Address the problem of even exposure of the commodity to heat and the possible problems of differentials in heat exposure.

Heat Treatment for Insect Control. http://dx.doi.org/10.1533/9780857097811.2.55

- Where in the production, transportation or storage process should the heat treatment be carried out?
- Link the heat treatment to the global pest management policy and procedures for the commodity from grower to consumer, addressing such issues as pre-treatment of storage and transportation facilities and prevention of re-infestation of treated commodities, complemented by ongoing monitoring and inspection.

There are a number of possible heat treatment techniques that can be applied to commodities, ranging from basic hot air heat treatments of boxes, sacks or tote bags in a kiln to complex microwave facilities. There are five main techniques we will consider here:

1. Forced hot air chambers and structures.
2. Fluidised beds and spouted bed systems.
3. Microwaves.
4. Vapour treatments.
5. Hot water dips and sprays.

6.2 Heat treatment techniques: hot air chambers and kilns

Forced hot air chambers (kilns) are most commonly used for timber and wooden pallets and packaging to meet both phytosanitary requirements and the requirements of ISPM15, but they can also be adapted for basic heat treatments of commodities in sacks, boxes or tins. Treatments can take 24 to 48 hours, or even longer for one tonne tote bags, bearing in mind that core temperatures need to reach a lethal 50 °C. The major potential problem is that a commodity close to the outside of the box or bag may be exposed to high temperatures of up to 65–75 °C for long periods of time. This is not normally an issue for commodities that will be cooked or roasted such as nuts or cocoa beans, but would be unacceptable for products such as grain for germination or wheat intended for bread flour, where intact protein or enzyme activity may be a specification requirement.

The other issue to consider here is water loss, both in terms of energy absorption to meet latent heat of evaporation requirements and the impact of water loss on weight and therefore value of the cargo. If air is recirculated, as it almost always is in fixed structures, any evaporated water will remain in the kiln and as the rH (relative humidity) of the recirculating air increases towards saturation so further rates of water loss are reduced. However, on completion of the holding phase of the heat treatment, temperatures will drop and this water will drop out of the air onto the surface of the commodity, which may then cause a mould problem once it is exposed to spores from outside.

One solution is to build in a period of forced ventilation, in which fresh air is circulated through the kiln to force out the water-laden air, normally with the heaters still on. This will prevent the formation of mould, however it will also result in the loss of the mass of water that has come out of the commodity. In Israel, work by Schlomo Navarro using temporary drying structures for control of *Carpophilus* sp. in dates using a solar drier for just three hours seems to have been very effective. Drying of commodities reduces the chances of reinfestation and reduces mould and bacterial activity, enhancing storage life.

6.3 Heat treatment techniques: fluidised beds and spouted bed systems

A fluidised bed is simply a sloping metal plate perforated with holes of a certain size and orientation, through which hot air flows up through the grain as it moves down the incline. The temperature and rate of flow are matched to expose the grain to a pre-determined temperature × time exposure period (CTP). Various suggested CTPs are 70 °C for 2 minutes or 60 °C for 12 minutes, 70 °C for 6 minutes or 80 °C for 4 minutes. Units have been created with capacities of up to 150 tonnes per hour. Even the normally obstinate Lesser Grain Borer (*Rhizopertha dominica*) has been reported to be killed at higher flow rates of 200 t/hr with grain temperatures at 65–70 °C and average treatment times of 2–4.5 mins (Qaisrani, 2000). CSIRO research shows grain flowing at 360–500 kg/hr at a depth of 100–250 mm across a fluidised bed with air inlet temp of 80–90 °C is completely disinfested within three minutes.

In spouted bed systems the heat is applied at the base of a silo and the grain at the top. Grain is forced vertically out of the centre of the silo through a vertical pipe where at the top the air is discharged through a filter and the grain drained off by gravity into an appropriate collecting vessel for further onward conveyance.

6.4 Heat treatment techniques: microwave systems and vapour treatments

There are a number of scientific papers describing the possibilities of treatment commodities with microwaves, in particular Wang *et al.* (2003) and Ahmed *et al.* (2011), which indicate that commercial plant developments are likely to be very expensive, but feasible.

Vapour treatment is a technique originally developed for the treatment of fruit fly infestation in fruit for phytosanitary purposes, whereby air saturated with water vapour at 40–50 °C was used to kill all life stages of the insects. The energy is transferred by the condensation of water onto the cooler surfaces of the fruit. It was originally used as far back as the 1930s to kill Mediterranean and Mexican fruit flies, but fell out of favour with the introduction of methyl bromide, ethylene di-bromide and ethylene oxide. However, with the decline in the use of fumigants for public health and environmental reasons, vapour treatments have come back into fashion, with modern units using forced air. This system has been used for subtropical fruits such as mango and papaya but studies have shown this treatment technique is treatment feasible for many fruits and vegetables.

As with normal hot air heat treatments there is a period of pre-warming (approach time), a holding time and then either natural air cooling or forced air cooling. There are numerous permutations depending on the insect species and most crucially control of damage to the product. There are no published specifications, due to the fact that there are so many uncontrollable variables to take into account, in particular the moisture content of commodities, fruit or flowers. Vapour treatments have been suggested for the control of insects in tropical cut flowers, but vase life may be affected adversely by the length of time/temperature exposure required for 100% mortality, as have hot water dips.

6.5 Heat treatment techniques: hot water dips

Hot water dips are simply as stated, but with highly controlled temperature × time exposures. They have been suggested for a number of scenarios. Hansen and colleagues (1991) demonstrated control of aphids in red ginger using a 5 minute dip in a 47 °C water bath, and Hara and co-workers (1993) achieved quarantine security against magnolia white scale on bird of paradise flowers by immersion in water at 49 °C for 10 minutes. It seems that just 49 °C for 5–12 minutes may be successful against a range of insects and plant pathogens. It has also been suggested that reconditioning in hot air at 39 °C for 2–4 hours may help reduce damage.

Gratwick and Southey (1985) report that hot water temperatures of around 40 °C have been used to treat bulbs, seeds and other planting material against disease; treatments lasted several hours and it can be concluded that heat treatment may have its place in the treatment of commodities and plant materials. In the animal kingdom, heat treatment kills by destroying proteins (including enzymes) and there are few examples of truly heat tolerant insects or animals. However, it should be remembered that in the plant kingdom there are grasses and other plants that only germinate or come to life after bushfires, so there is a significant amount of potential research to be done in this field. If appropriate precautions are taken to keep produce clean, there is no reason why, for example, cocoa beans cannot be heat treated early on in the transportation process, kept sterile (i.e. not re-infested) and go all the way to the chocolate factories without the need for further treatment, resulting in significant cost savings. The same principle may apply to a whole range of commodities.

References and further reading

Ahmed, M. *et al.* (2011), Disinfestation of *Rhyzopertha dominica* using microwave heat treatment to the Malaysian paddy. *World Academy of Science Engineering and Technology* 57.

Bell, C. (2014), 'Pest control of stored food products: insects and mites', in Lelieveld, H., Holah, J. and Napper, D. (Eds.), *Hygiene in food processing: principles and practice* (Second edition), Woodhead Publishing Limited, Cambridge, UK.

Gratwick , M. and Southey, J.F. (1985), Hot water treatment of plant material. *Min. Agric. Fish Food*, London, 64 pp.

Hansen , J. D., Hara, A. H., and Tenbrink, V. L. (1991), 'Recent progress in the control of insect pests on tropical floral commodities', in Leonhardt, K. W., Evans, D. O. and Halloran, J. M. (Eds.), Univ. Hawaii Res. Exten. Ser. 124, pp. 54–60.

Hara, A. H., Hata, T. Y., Hu, B. K. S. and Tenbrink, V. L. (1993), Hot water immersion: a potential quarantine treatment against an armored scale *Pseudaulacaspis cockerelli* (Cooley). *J. Econ. Entomol.* 86, 1167–1170.

Heaps, J. (Ed.) (2006), *Insect management for food storage and processing* (Second edition), AACC Publishers, USA.

Hill, D. (2002), *Pests of stored foodstuffs and their control*, Springer Publishers, The Netherlands.

Kloosterman, L. and Mager, K. (2014), 'Pest control in food businesses: an introduction', in Lelieveld, H., Holah, J. and Napper, D. (Eds.), *Hygiene in food processing: principles and practice* (Second edition), Woodhead Publishing Limited, Cambridge, UK.

Lurie, S. (1998), *Postharvest Biology and Technology* 14014, pp. 257–269.

Morland, D. and Hedoes, S. (Eds.) (2004), *The Mallis handbook of pest control* (Ninth edition), GIE Media Inc., USA.

Qaisrani, R. (2000), *Farming Ahead* no. 106, Oct 2000, CSIRO.

Subramanvam, B. (Ed.) (1995), *Integrated management of insects in stored products*, CRC Press, Boca Raton, USA.

Wang, S., Tang, J., Cavalieri, R. P. and Davis, D. C. (2003), Differential heating of insects in dried nuts and fruits associated with radio frequency and microwave treatments. *American Society of Agricultural Engineers* 46(4): 1175–1182 E.

Part Three

Other applications

Heat treatment of insect pests in buildings

7

7.1 Introduction

Heat treatment of whole buildings has gained in popularity with many clients, particularly in the flour milling industries where this is seen as an alternative to fumigation with methyl bromide or other fumigants. However, before embarking on an expensive heat treatment, the owner should consider where the insects are breeding and whether it is cheaper to just treat the machinery either in one go or in a rolling treatment programme. Insects do not live in open space and it could be considered a waste of energy to heat up thousands of cubic metres of air when in fact the insects are confined to a few tens of cubic metres of machinery.

There may be other complications with full space treatments, namely how to deal with sprinkler systems and the insurance implications of draining sprinkler pipes and removing or replacing heads. Some machinery such as plan sifters may use glue to secure the sieve meshes in place, which can be affected by heat and so should be protected.

Having decided to carry out a space treatment, the first question for the surveyor concerns the number of heaters required. When installing a new heating system in a building, a structural heating engineer may take a couple of days to survey the building and work out the exact surface areas of different types of material coating the outside of the building (types of bricks, insulation, surface area of glass, the type of glass and whether double glazed or triple glazed, types of roofing material, etc.) in order to work out the potential heat loss. He will have an idea of the mass of the building and from all this deduce the size of boiler to install and where to locate radiators for maximum effect and efficiency.

The heating techniques used are, as for rooms:

- convection systems
- hydronic systems
- electric heaters.

Direct burn gas systems should not be used due to the risk of combustion, CO and CO_2 build-up and introduction of water into buildings due to the gas combustion process. On the other hand, convection systems have a number of advantages. They take fresh air from outside, and force it into the building, displacing existing cold air along a selected exit route – normally but not necessarily high up (Figure 7.1). Heating causes evaporation of water from stock, machinery and building fabric, with a beneficial effect on future mould growth.

Heat Treatment for Insect Control. http://dx.doi.org/10.1533/9780857097811.3.63

Figure 7.1 Heat treatment of buildings. (Top photograph courtesy of Benedikt Segur)

In the case of flour mills, it also improves flour flow along pipes – the drier the flour stock, the more easily it flows. As the production rate of a flour mill is determined by the maximum rate of flow of flour at its most constricted point, even increasing rate of flow by 0.5% by drying out the process may have significant production benefits. A dry, warm, well exhausted flour mill will work more efficiently than one clogged up with damp, mouldy insect-infested flour residues in pipes and elevators.

For each system, the following stages apply:

1. Survey, risk assessment and planning.
2. Building preparation.
3. Pre-treatment with insecticide (non-organic sites).
4. Set up equipment and temperature monitoring systems.
5. Heat treatment.
6. Tidy up.

7.2 Survey, preparation and pre-treatment

The survey is crucial and needs to answer a number of key questions. The size of the task in hand is important and the surveyor needs to determine not only the volume of the treatment area but also the approximate mass of the building. At this stage a decision also needs to be made with the customer as to whether or not the heat treatment may be supplemented with residual insecticidal treatments that may significantly enhance the effectiveness of the treatment and also reduce the treatment time and resources required for the job.

From a risk assessment and health and safety point of view, the surveyor will look for fire and heat detectors, sprinkler systems, compressed air vessels including fire extinguishers, compressed air tanks and even aerosols including deodorant cans in private houses or hotel rooms. These will need to be drained, emptied, disabled or removed as required.

The other important points to identify at an early stage are items which may be adversely affected by heat; for example in flour mills, plan sifter sieves, high protein ingredients (proteins denature when exposed to heat) and live materials such as yeasts are at risk. In hotels and homes, paper products and painting, any furniture containing some glue products such as veneers and even ornamental features stuck on with glue may be affected at higher temperatures.

At higher temperatures there may be some temporary structural warping of door frames, drawers may not move easily and where metal features or equipment are bolted into wooden bases small cracks may develop where the metal expands more than the wood. The amount and nature of any potential damage is dependent on:

- the product of the maximum temperature and time;
- the moisture content of the wood prior to treatment;
- the fragility of the item at risk.

From the outset there is a conflict of commercial requirements versus technical ideals. A rapid, high temperature treatment of between 60–70 °C will get the job done more quickly in a building, but will also significantly increase the risk of damage to the areas mentioned above. However, commercial pressures to get hotel rooms back into service, or production back up and running, sometimes dictate that rapid treatments be attempted. Damage is often minimal or manageable. However it is most important that the operator in charge points out possible areas of damage to the client and agrees where the acceptable limits lie.

It is worth remembering that unlike fumigation, heat treatment allows for parts of the building to be accessible at all times. Taking the necessary precautions into account, it is therefore possible for the client to be on site during heat treatment to observe any potential areas of concern and address any problems, e.g. by adjusting the temperature or treatment time.

For heat treatments in buildings, water may be introduced unintentionally into the atmosphere by such factors as the drying effect of heat on a damp structure, from water in a pre-treatment insecticidal spray or from open water sources such as toilets, sprinkler tanks or pre-treatment washing of floors. Water can have a huge effect on heat treatment.

One reason for this is that the movement of water between air and wood as measured by the relative humidity (rH) of water in air versus the moisture content (%) of water in wood determines the level and effect of warping of wood and timber. Rapid heating of one side of a damp timber plank or floorboard using dry air will cause water to evaporate more quickly from the exposed side than the other, warping the board. For example, on a floor this will be the top exposed surface, causing the board to bend (warp) upwards.

Some excellent work by the late Werner Von Rotberg of Thermo-lignum has created a system based on the psychrometric charts, which is used to heat up and cool down wooden museum artefacts slowly by matching the rH of the air to the moisture content of the wood, so that there is no movement of water either from wood to air, or from air to water. However, as the temperature of the air increases during the heat treatment, so its rH decreases. A computer program is used to add water to the air to keep the rH at the right level to the moisture content of the wood while at the same time taking account of the temperature of the wood. It does this automatically not only during the heating phase but also crucially during the cooling down phase where water is taken out of the air to prevent condensation and adsorption of water into the wood, the net result being a perfect heat treatment with no damage to wood due to warping.

A second effect is that water is an excellent coolant due to the large amounts of energy required to evaporate it. This can be fatal for heat treatments where losing heat is generally the last thing required. Outdoor treatments in temperate climates can be made impossible by rain and an open water source in a bakery cooler can have the same effect as putting an ice cube in your coffee: rapid cooling until the ice has melted, or in the case of the cooler until the water has evaporated. In hotel rooms, the porcelain bowls of toilets may heat up considerably and cause water to evaporate from the toilet into the room, causing damp problems when the room is cooled down. Opening windows may allow water-laden air to escape, provided it does not cool down too quickly and drop its load of water by condensation onto wallpaper, etc. The answer is to allow any cleaning water to dry prior to treatment and cover all exposed water sources, including toilet bowls.

The relationship between water and heat is also affected by the type of heat treatment used. Direct introduction of dry air convection systems from heat exchange units will, unless recirculated, evaporate water and push it out of the building. This makes such systems ideal for drying out buildings and for flour mills and other 'dry' buildings. Hydronic, steam radiator and electric systems tend to recirculate the air in the building and therefore if water is evaporated out of the structure, there is a risk of condensation on cooling unless separate fans and ducting are used to force water-laden air out of the building. Direct burn propane systems (not recommended) introduce water and CO_2 into the building, so apart from the obvious fume hazard, huge amounts of water are evaporated into the air, which will condense out on cooling.

Although clearly not an option for organic sites, supplementary insecticidal treatments can greatly enhance the effectiveness of a heat treatment. Indeed, just heating up a building to 40 °C following a residual spray treatment to the fabric of the building will flush insects out over the treated surfaces, which may be particularly effective against mobile insects such as cockroaches and bedbugs for short term effect. However, eggs and pupae (in the case of complete metamorphosis insects) will not be affected by this kind of combined treatment, but require heat to penetrate machinery and cracks in wall/floor junctions, etc.

Another important consideration for the survey is air movement. Still air is an insulator; if there is no forced air movement then eventually small natural convection currents will develop as lighter, less dense hot air moves upwards and displaces cooler air. However, in low corners or inside machinery there may be no air movement at all. Here convection systems with powerful fans can be more effective than hydronic, radiator or electric systems. Using convection systems may be at the expense of lower air temperatures, increasing treatment times, but this can be beneficial in terms of reducing damage.

The next consideration is the type and layout of the building. For large hotel jobs where the contractor may be treating over 40 rooms at a time, it is almost impossible to set up a viable convection system to blow air simultaneously into 40 or more different rooms. It is, however, quite straightforward to put either a small heater or small radiator unit in each room connected back via high pressure pipes to a central trailer-mounted boiler unit.

Insulation is a major consideration, along with the decision as to whether to attempt to heat the whole structure of the building up to a lethal 50 °C temperature or to limit the core temperature targets to machinery and floor areas. As for any pest control treatment, the real question that needs to be answered in this case is: where are the target insects harbouring? In warmer climates, where buildings are often not insulated properly, it may be necessary to insulate roof areas or even walls.

If the building to be treated is a well-maintained modern flour mill where insect infestation is limited to the insides of modern inaccessible machinery and bins, not in the building fabric, then the choice is between spot treatment of the machinery and heating up the building and spraying the fabric to contain any insect eruptions, but not trying to heat walls or concrete structures. This allows for a much shorter heat treatment, and as well as minimising the risk of heat damage and reducing disruption to production, requires in temperate and northern climates (where mills are insulated) almost no insulation other than shutting doors and windows. Small spot treatments can be carried out regularly as part of a planned maintenance/hygiene programme and have the advantage of offering year round insect control opions, whereas a single annual heat treatment has the same problems associated with it as did annual space fumigations, and that is the problem of high insect numbers and risk of product contamination in the months leading up to the treatment.

Conversely, an old wooden organic mill with a fabric ridden with flour beetle and mill moth will require heating slowly and gently to the core over 24 to 48 hours with the associated major disruption to production schedules.

7.3 Preparation of the building

The survey will have determined the equipment required, the level and location of infestation, the insulation required (if any) and how long preparation and setting up will take the contractor. However, the client will also need to do a significant amount of preparation.

Fire control and suppression systems are obviously heat sensitive and need to be made safe. Sprinkler systems are another obvious risk. Different types are activated by different temperatures. Table 7.1 shows NFPA 13 table 3-2.5.1 (US National Fire Code), which defines the different sprinkler head ratings. If heat treatment is to be a regular insect control option, then sprinkler heads should be changed for as high a rating as possible while still meeting the requirements of fire control officers and

Table 7.1 **Sprinkler head ratings as defined by NFPA 13 table 3-2.5.1 (US National Fire Code)**

Classification	Maximum ceiling temperature	Sprinkler activation temperature	Glass bulb color	Fusible link color
Ordinary	100 °F	135 °F–170 °F	Orange (135 °F) Red (155 °F)	Black; no color
Intermediate	150 °F	175 °F–225 °F	Yellow (175 °F) Green (200 °F)	White
High	225 °F	250 °F–300 °F	Blue	Blue
Extra high	300 °F	325 °F–375 °F	Purple	Red
Very extra high	375 °F	400 °F–475 °F	Black	Green
Ultra high	475 °F	500 °F–575 °F	Black	Orange
Ultra high	625 °F	650 °F	Black	Orange

Source: Reprinted from http://archtoolbox.com/materials-systems/fire-supression/85-sprinklerheadtypes.html, accessed 25.02.14.

insurers, although insurers will also consider the potential damage caused by activation of heads by heat.

Ideally, therefore, sprinkler heads should be of as high a rating as possible. They may then be protected from heat by placing a damp tissue on each head, covered with a plastic cup and a polythene bag – the highest risk ones may then also have a temperature sensor placed under the protective covers for monitoring during the course of the heat treatment. Whole sprinkler systems may be drained – again by agreement with fire control officers and insurers – as an additional precaution.

Pressurised vessels containing fire suppressant materials must be protected from heat or de-pressurised, and any heat sensors deactivated, again in consultation with fire control officers and insurers.

One of the major reasons for failure of any heat treatment is the failure to locate and identify significant deep residues of food stock, whether this is stock dropping from the walls of silos into the silo bottoms or simply infested stock left inside machinery. For all heat treatments of mills, etc., all stock and bulk commodities that may be infested should be cleaned out, all machinery run empty and ideally cleaned out as thoroughly as possible, all areas of the fabric of the building cleaned and sprayed where appropriate and as many bins as possible emptied and the sides banged to dislodge adhering stock, then run clean so that no residue remains in the bottoms.

The stock in those bins that remain full is likely to remain unaffected by the heat, particularly in large bins. Small amounts of high protein flour close to the sides of bins may be denatured, but when delivered and mixed with the content in the central areas of the bin the total performance of the flour is unlikely to be seriously compromised. However, it is always worth testing prior to delivery as a precaution.

When treating hotels or private houses all clothing must be hung up and spaced out and all bedding and mattresses separated and stacked in such a way as to allow free air circulation. Drawers full of clothing will not allow for heat penetration, meaning that survival of insects is likely.

7.4 Setting up of equipment, heat treatment and temperature monitoring

Whichever heat delivery system has been chosen, the objective is even delivery of heat to all areas, reducing extreme cold or hot spots to a minimum and ensuring that target temperatures are achieved for the desired lengths of time in the actual areas where the insects are living.

It is totally unacceptable just to measure air temperatures as an indicator of the performance of a heat treatment. Temperature sensors must be located to give the operator in charge the following information:

- the distribution of heat throughout the building and relative rates of heat increase;
- identification of the hottest and coldest parts of the job, including inside machinery, bins and silos, wall voids, sub-floor areas, in mattresses, and risk areas such as delicate electronic equipment or sprinkler heads;

- to identify when each area has reached its temperature × time target so that heat resources can be re-deployed to those areas that are struggling to reach target.

Proper data logging with instant graph plotting is crucial to giving the operator in charge the information he needs to manage the heat treatment and to provide a record of treatment for the client.

Temperature monitoring systems may be cable linked or radio linked. Cable systems are generally more reliable and not subject to interference, however they may be less practical in large buildings or large, multi-roomed hotels.

7.5 Heat treatment

The quality of heat treatments for buildings varies massively around the globe, from very high temperature operations either by direct burn heaters or by those more used to drying out damaged buildings, to refined, computer-controlled systems regulating temperature and humidity in each area to prevent any risk of damage.

The choice of heat treatment type and contractor is often price-driven, but this can be a false economy. For many years heat was vetoed as a possible replacement for methyl bromide fumigation because of stories of horrendous damage, sometimes even serious structural damage, caused by irresponsible heat treatment operatives. It is quite unacceptable for heat treatment technicians to cause damage to any building or piece of equipment, and quite possible for a competent surveyor to predict risk areas with a degree of certainty and then take appropriate measures to minimise those risks.

The first step from a heat treatment contractor's point of view is to agree the acceptable temperature range with the client. One example was the treatment of an aircraft, rated in its manual to operate up to 65 °C or sometimes more. The heat treatment operative in this case agreed with the customer to keep the maximum air temperature below 60 °C and to aim for core target temperatures of 50 °C to 55 °C. Given these unequivocal parameters, and potentially expensive repercussions for stepping outside the limits should any damage occur, there was no option but to keep the treatment within these parameters. The location of the temperature sensors was agreed with the client, and a record kept from each temperature sensor. By doing this the balance of risk was left with the client, and a safe and effective treatment was carried out.

It is widely accepted that damage to facilities may start to occur above 65 °C, and for this reason it is strongly recommended that heat treatment technicians aim for air core or surface temperatures of between 50 °C and 60 °C resulting from air input temperatures of 60–70 °C, never exceeding 70 °C in any risk situation. Even in ISPM15 kilns where the target temperature in the core of thick, damp timber needs to reach 56 °C, it is recognised that a 70 °C air temperature is sufficient to achieve this within a reasonable time period (a few hours, depending on thickness of timber and start temperature). Raising air temperatures in buildings to 75 °C or higher is potentially highly risky, and in the opinion of the author should not be attempted. Of course, whether or not damage occurs at these temperatures depends entirely on the contents of the building and the building structure. However the risks do not outweigh the benefits in terms of insect control, except possibly if one were aiming for some form of emergency *Salmonella* control, which is outside the scope of this book.

There are a number of ways in which buildings can be heat treated, but unlike fumigation it is not technically necessary to achieve the desired temperatures throughout the whole building at the same time. If resources are limited, there is no reason why a large building should not be treated in sections, with waste air from one section being vented on to pre-heat the next section. Indeed, when treating multi-storey flour mills with convection systems it is conventional to start by putting the bulk of the air input resources at ground level, venting out the roof or top floor, either into the atmosphere or by recirculating the hot air back into the heating units.

Once the ground floor has reached its temperature × time target, the input ducting can be moved to the first floor, which has already been pre-heated by the rising air from below, and so on upwards until the job has been completed. Using such techniques, smaller heating units can be deployed to heat quite large structures.

Hydronic systems deploy a central, trailer-mounted boiler linked by hydraulic hoses to a large number of radiator units deployed around the building. The energy is transferred quickly from the boiler to the radiators by a heated water/glycol mixture, which is then recirculated back to the heater when it has given up its energy to the radiators. The equipment is expensive and does take quite a while to set up, but can prove very effective.

Another alternative that has proved popular in some areas is the installation of large numbers of 18–19 kW electric heaters. These are generally three-phase 415 volt units linked to the client's power supply. They have their place, but require a qualified electrician to set up and operate safely and the risks of high voltage/high current cables linked to possible sprinkler activation should be addressed. They are a clean and relatively easy option to deploy. However poor fan performance, large client electricity bills and lack of power have led to failed treatments. A single 450 kw convection heater or hydronic system delivers the equivalent power of around 25 18 kW electric heaters. As stated previously, direct burn propane heaters should not be used due to the risk of collateral damage.

7.6 Finishing the treatment

The timing of the decision to end a heat treatment is crucial. Insects are only killed in the last hour or so of the treatment and for much of the treatment there are no insect casualties. In fact, the heating process takes insects through their peak performance zones. Proper monitoring of correctly placed temperature sensors gives the operator in charge an idea of when to stop the treatment. But before turning off the heat a final visual check needs to be carried out for any live insects, and a heat camera or infra-red thermometer may be deployed at any time to check for any overlooked potential cold spots.

It is vital in food factories that the client builds-in time for cleaning before commencing production. Insects previously harbouring in small spaces will have died in their thousands on food-contact and food-flow surfaces, and these will end up in the next production run unless cleaned up or flushed through.

Sometimes heat treatment technicians will have secured temperature sensors inside machinery with tape – it is vital that this is not left inside as it is a potential contaminant. Metallic tape is recommended if this has to be done, as it can be detected using a metal detector, and a strict policy of one piece in, one out adopted. It is not recommended to secure a sensor with more than one piece of metallic tape, as the person who removes the sensor may not be the same person who installed it.

Finally, before leaving the site the operator in charge must always walk around with the client to check nothing has been left in a dangerous condition; for example hatches opened or production settings inadvertently changed on machinery.

References and further reading

Bell, C. (2014), 'Pest control of stored food products: insects and mites', in Lelieveld, H., Holah, J. and Napper, D. (Eds.), *Hygiene in food processing: principles and practice* (Second edition), Woodhead Publishing Limited, Cambridge, UK.

Heaps, J. (Ed.) (2006), *Insect management for food storage and processing* (Second edition), AACC Publishers, USA.

Hill, D. (2002), *Pests of stored foodstuffs and their control*, Springer Publishers, The Netherlands.

Kloosterman, L. and Mager, K. (2014), 'Pest control in food businesses: an introduction', in Lelieveld, H., Holah, J. and Napper, D. (Eds.), *Hygiene in food processing: principles and practice* (Second edition), Woodhead Publishing Limited, Cambridge, UK.

Morland, D. and Hedoes, S. (Eds.) (2004), *The Mallis handbook of pest control* (Ninth edition), GIE Media Inc., USA.

National Fire Prevention Association (2013), NFPA 13: Standard for the Installation of Sprinkler Systems.

Subramanvam, B. (Ed.) (1995), *Integrated management of insects in stored products*, CRC Press, Boca Raton, USA.

Heat treatment of bedbugs and other insect pests in small rooms

8

8.1 Introduction

Insects, particularly bedbugs, will easily survive in drawers full of clothing, inside suitcases and even in thick piles of dust in a vacuum cleaner. As a room heats up, mobile insects such as bedbugs and cockroaches will actively and quickly seek out cooler refuges in cable conduits, drains, under floorboards or lie against cool glass or water pipes to effect survival. In 99% of cases, excepting the most simple cell structure rooms, a full survey is required to ascertain the extent of an infestation and alternative measures taken to deal with insects in these 'refuge' areas.

The technical aspects of heat treating a room are relatively simple. Heat should ideally be directed into the corners of the room, with air being forced back towards the door and re-circulated to the corners (Figure 8.1). The structure is uncomplicated in terms of air circulation and most rooms are not large compared to a large building or silo, so loss of energy is rarely an issue. In addition, if the heaters are placed inside, no restrictions are placed on the job by the weather. However, these often apparently simple jobs can be fraught with pitfalls and claims with regard to damage to property (see Figure 8.2) and failed treatments.

8.2 Assessing the problem

As with all heat treatments, the preliminary survey is most important. The following list of general questions should be clearly answered by the client. Domestic householders or hotel staff may not have any understanding of pest control, unlike food clients who are often very experienced due to food quality and safety requirements.

8.2.1 What is the pest problem species and where did it come from?

If bedbugs have been found, the pest control operator (PCO) should be sure they are not pigeon or bat bugs. If fleas are the problem, treatment should be carried out in conjunction with treatment of the infected animals and good cleaning and sanitation of any baskets, cages or habitual resting/sleeping areas of the pet. Identification of cockroach species is crucial, as it will determine whether heat treatment is possible or not. Oriental cockroaches *Blatta orientalis*, often breed in the bowels of a building and around drains which will be inaccessible to heat penetration, whereas German

Heat Treatment for Insect Control. http://dx.doi.org/10.1533/9780857097811.3.73

Figure 8.1 Heat treatment of a room. (Photograph by Steve Kershaw)

cockroaches *Blattella germanica* need heat to survive and will have populations concentrated around heat sources like backs of fridges and freezers or boilers or ovens. These localised populations may be isolated and heat treatable.

8.2.2 Where exactly is the infestation and where could it have spread to?

Bedbugs usually remain close to the food supply for feeding, however females may move away from feeding areas such as beds and sofas to lay their eggs. Bedbugs can migrate large distances relatively quickly and survive for long periods without food. If rooms have been left empty (e.g. hotel rooms thought to be infested) and bed bugs have gone for long periods without a feed, they will go looking for a blood meal elsewhere and may be found in other parts of the building.

8.2.3 If a single room is heated (e.g. an isolated hotel room) what possible options are there for insects (e.g. bedbugs or cockroaches) to escape or avoid the heat?

Cable conduits, sub-floor areas, drains and ceiling voids are all possible escape routes. These need to be sealed, treated with insecticidal dusts (including desiccant dusts) or spray, or even sealed with sticky pads or other traps to prevent migration.

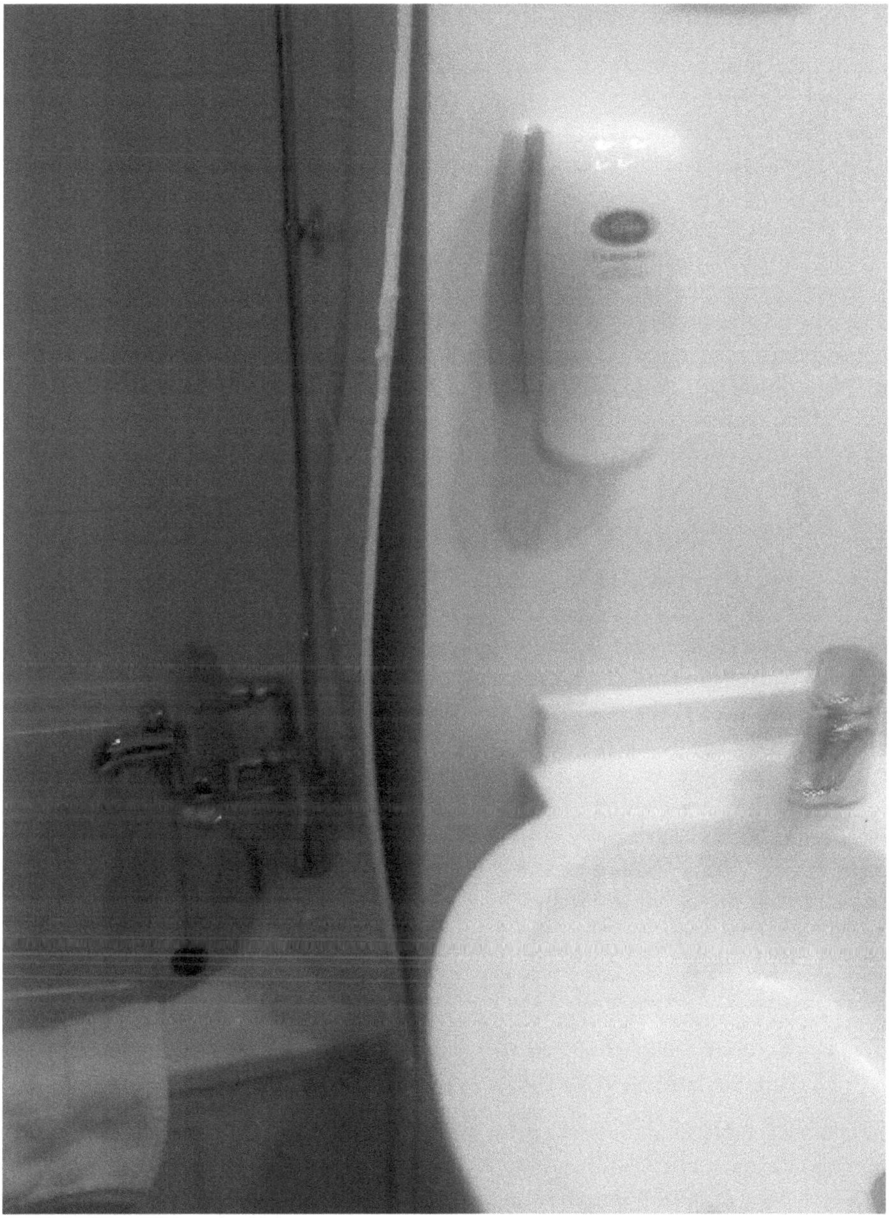

Figure 8.2 Damage to room fittings caused by heat treatment. (Photograph by Steve Kershaw)

8.2.3.1 Case study: cockroach treatment in kitchen drains

During the heat treatment of a kitchen facility against cockroach infestation, the operator started heating up the kitchen and was quickly called outside to observe hundreds of cockroaches pouring out of the drains. It was not practical to spray insecticide down the sewers or to heat up the drains to any lethal level. In the end, thinking outside the box, the operator put a ring of residual insecticide on the floor around each drain in the kitchen and blew heat from the outside back up each drain, thus driving any cockroaches trapped in the tunnels back into the kitchen and over the treated surfaces, and then completed the kitchen treatment. With regard to the question on the original survey list, 'If a single room is heated (e.g. an isolated hotel room) what possible options are there for insects (e.g. bedbugs) to escape or avoid the heat?' this cockroach story is a classic example of the sort of potential pitfall that needs to be overcome using a more holistic approach to a pest control problem.

8.2.4 What is the potential for damage if too much heat is applied?

Is the customer willing to either accept the risks or to be continually present during the heat treatment to review progress and monitor any slight movements in wooden or timber elements or other risks that may mean treatment needs to be halted?

The survey will reveal the presence of risk items such as paintings, tins of food or drink, fire extinguishers, lighters, aerosols, etc., which should all be removed. Ornate plaster ceilings and cornices should be regularly checked throughout the course of the treatment for any signs of cracks, etc., brought on by drying. This can be prevented by humidifying the air during the heating phase and venting air to the outside during the cooling phase to prevent condensation. However, too much humidity caused by pre-spraying with a water based insecticide then heat treating with internal heaters can humidify the air so that water is absorbed into wooden floors and cause warping. Close monitoring and an understanding of the rH of the air during treatment is strongly recommended.

8.2.5 Is heat treatment alone likely to solve the problem or are supplementary insecticidal treatments also required?

Is this acceptable to the client (many private individuals will have opted for heat to avoid chemicals)? If so, what are the risks to children or pets? From a technical heat treatment control point of view, using the full range of appropriate commercial pesticides to supplement the heat treatment will achieve the best results. However, many clients may wish to limit or even totally eliminate exposure of themselves and their children to potentially harmful pesticides. In these cases, the potential limitations of the various options should be explained to the client, who must co-operate fully by doing their bit in helping make the treatment a success.

Unless a small-scale infestation really is limited to one sealed room, it is almost totally impossible for a pest control operator to simply carry out heat treatment of one room in isolation and expect 100% success. In these cases, the entire house (or indeed hotel) should be heat treated, which will in most cases take upwards of 16–24 hours for the heat to penetrate properly into walls and wall cavities. This is not a cheap operation. The client will need to hot wash or deep freeze all at-risk items such as clothing, towels and bedding, or spread/hang them so they are exposed to the heat for the full time during treatment. Children's toys, books and other household goods may harbour insects and should be spread out to ensure good circulation of the hot air. Heat treatment operators should have a large range of fixed sensors to monitor the gradual increase in core temperatures, there may be over 100 sensors (e.g. Thermolog radio sensors) positioned in potential cold spots, e.g. down the backs of sofas, inside bed mattresses and quilts, behind wall cabinets, inside drawers, inside the backs of white goods and inside accessible wall cavities or dry risers throughout the full range of the house or hotel. These fixed temperature sensors may be supplemented by infra-red spot meters or thermal imaging cameras, but such equipment is no substitute for fixed-location sensors, which are the only true arbiters of the success or otherwise of the heat treatment.

8.2.6 Are there any smoke detectors or heat detectors that may be affected?

Smoke detectors work by detecting fine particles of smoke and may be confused by dust particles being thrown into the air by the increased air movement required for successful heat treatment. These can be covered to prevent accidental activation. Heat detectors are normally electronic instruments linked to a central fire control panel and may be isolated at this central panel. It is often an insurance requirement that active fire monitoring systems be in place at all times – it is good procedure to advise the client to consult with his insurers should the fire monitoring systems be isolated. Having said this, these automatic systems will be replaced by having a man on hand at all times to monitor any problems. Heat treatments should never be left unattended for any significant length of time.

8.2.7 Is the client fully aware of the implications of the treatment and, if only treating one room instead of the whole building, aware of the possibility of re-infestation from adjacent non-treated areas?

There is no way that a single survey can determine for certain that any infestation is confined only to one room or one area. The surveyor can only say that he cannot find any evidence of infestation elsewhere, and evidence in the form of droppings, smears or actual insects or insect parts requires time to develop. If the client wants to save money by only treating one room, treatment should proceed at the client's own risk and this should be explained to him at the time of the survey. The survey will also include the usual components of a pest control survey: an outline of how the job will actually be done and risk assessments, etc.

8.3 Treating bedbugs

While some bedbug infestations necessitate whole building heat treatment, in a significant number of cases either the client cannot afford this or the small size and localised nature of the infestation does not warrant the expense or risk of damage to the building fabric, structure or contents. Under these circumstances, combination heat treatments have proven very effective over the years. In such cases, "at risk" furniture and fittings are heat treated either under sheets or in a heat treatment "bubble" and the room fabric sprayed and/or dusted with residual insecticide.

As with almost all pest control treatments, the procedure starts with a detailed survey to establish the full extent of the bedbug infestation. Sniffer dogs and electronic devices may be used to supplement inspection and service reports. A standard procedure (based on a 2003 Thermokil operating manual) is:

The room first needs to be prepared for treatment:

1. All bedding, curtains and covers should be bagged up and sealed, and removed for immediate laundering at 60 °C or more.
2. Cover up all smoke alarms and heat detectors, get hotel management to notify the fire brigade if an automated fire alarm system is in use.
3. All furniture should be moved away from walls, and headboards, etc, or physically taken off the walls. The client should do this, as often there are electrical cables that will need isolating. Plug sockets should also be dropped down.
4. The whole room should be treated with a suitable residual insecticide – Tenopa is often considered a good choice.
5. A small heat resistant groundsheet should be placed on the floor and treated both sides with residual insecticide.
6. The furniture should all be stacked on the groundsheet to form an inverted U shape, with the mattress(es) lying across the top to form the cap of a central area into which the hot air will be blown. Where possible keep as little as possible of the furniture in contact with the floor, but still make the inverted U as airtight as possible. A bed base, raised off the floor slightly on two or three market clips and placed on its side makes a good back wall for your stack.
7. Often the two sides will consist of dressing tables or bedside tables, but small chairs and other miscellaneous furniture can be placed under the mattress(es) which form the roof of your stack.
8. Your stack should be constructed so that air is forced into the middle, has nowhere to go, and comes back up the sides. When we put the top sheet on we leave two vents in opposite corners at the back, the air is then forced down the outside of the stack and out the rear vents.
9. You should work to keep as much surface area of furniture exposed to hot air as possible. Tightly stacking furniture and mattresses just creates thick insulated layers in which insects will survive.
10. It is not recommended that televisions are placed in the stack, and remove all food and drink from the room – bottles of fizzy water may blow in extreme heat.
11. Some insulated areas are inevitable, and it is these areas we want to identify and to place our internal temperature sensors in. Typical cold spots will be where mattresses rest on top of rear wall beds, and where they come into contact with side-supporting furniture. Place temperature sensors right between the mattress and the bed base.

12. You may wish to make a small hole in the bed base to place a temperature sensor in. Also inside drawers and under cupboards are good places.
13. Always use a minimum of six temperature sensors on each job, placed in the following areas:
    T1 In open air – air input temp
    T2 Cold spot front left side (low between mattress and floor)
    T3 Cold spot back left side (high between mattresses lying across each other)
    T4 Cold spot back right side (low between mattress and floor)
    T5 Cold spot front right side (high between mattresses lying across each other)
    T6 In open air – air outlet vent
14. The top sheet can be put over the whole stack and tightly rolled and clipped to the base sheet.
15. Leave two open vents or "air drains" in each rear corner. The vents combined cross section area must be greater than that of the input duct, and increasing or decreasing the size of either vent will influence the route the hot air takes within the stack, i.e. increasing the size of exit hole is used to increase the temperature or flow of energy down that side.
16. If an internal heater is used, the exhaust may be vented either out of the room window or via the end-of-corridor fire exit. NB Trip hazard.
17. Make sure all cables are secured to avoid trip hazard and always use circuit breakers on wall sockets at beginning of cable runs.
18. Always fully uncoil extension cables.
19. Unit may then be switched on.
20. Target temperatures for bedbugs are different to other jobs because of the nature of the substrate in which they live and the difficulties of identifying the coldest cold spot. We must therefore achieve 52 °C for a minimum of two hours in all areas before taking off the top sheet. A typical single or double room will take three to four hours.

Occasionally you may be able to do the furniture of four rooms in one job. The stacking regime may be different here. We have found that by stacking beds horizontally on top of each other, being spaced apart by their legs or other spacers, and placing the mattresses on their ends around the stack of beds, with the headboards, etc., on the outside of the mattresses, the job can normally be done in six to seven hours. The otherwise empty rooms can then be treated during the course of the heat treatment. Using this technique, and assuming access for ducting from outside (ie rooms no more than 8 m off the ground with windows that open), then up to eight rooms can be completed in a normal day using a typical double heater kit.

8.4 Types of heat treatment

If the perimeter of an infestation has been clearly defined, whether it is cockroaches, bedbugs, fleas or clothes moths, measures should always be taken to make any possible escape routes toxic by the use of insecticidal dusts or sprays before the heat treatment is carried out. Rooms may be treated in a number of ways:

- Hot air may be ducted into the room from a diesel heater either outside in the corridor or in a vehicle outside. In this case, air is either applied from the door and vented out the window or heat may be ducted into the far corners via a plenum box and blown back out the door. Air may also be ducted straight into cupboards and wardrobes with a separate supply into any en-suite bathroom. The advantages of this system are that more vigorous air movement and powerful directed flow is good at penetrating less accessible voids. The disadvantage comes if a large number of rooms require treatment together, which requires a lot of ducting and a lot of air to move.

- Large numbers of rooms are better treated using either small-exhaust dried heaters, 18 kw three-phase electric heaters or a hydronic system, in which a radiator with a small fan is put in the room and heat fed from a powerful central boiler to each radiator in the form of a water and glycol mix via high pressure hoses. As there is less powerful convection in this case in each room, air temperatures are often increased to achieve the same level of heat penetration, which can increase the risk of damage. Conversely, because the same air is just being moved around the room (unlike the repeated full air changes of a convection system), the relative humidity (rH) of the air is higher – if water-based insecticidal sprays have been used at the same time the rH can jump quite high causing swelling of timber, as opposed to the drying effect of a convection system. Either method can cause temporary warping, making doors difficult to shut and drawers difficult to open. However, normality is usually restored once the rooms have cooled down.
- In the USA, some companies have been known to use direct burn propane burners. The use of such equipment is not recommended and should be avoided. There have been cases of houses catching fire due to propane burner use, and the high direct temperatures make the risk of damage much higher, not to mention hazards from fumes and the input of water from burning propane.
- Electric heaters are clean, easy to use and position, but expensive to buy and require large amounts of electricity (normally 3-phase) which is not usually easily obtainable in a hotel room. Thirty heaters require a lot of cable and a big generator, but this is not impossible by any means. Feeds can be ducted and split for better direction of air. Some pest control operators use low power electric heaters to re-circulate air rather than the diesel powered Thermokil systems. However, whichever commercial system is used, the principle remains the same: heat the furniture, spray or dust the fabric of the rooms and hot wash all bedding, clothing and curtains.

8.5 Ensuring effective treatment

After the means of heating the room has been decided on, elimination of insect refuges is important, together with an understanding of the location of cold spots and how to measure them. Cold spots are found amongst piles of clothing, bedding or sheets, which must be stacked or hung separately so that hot air can circulate freely between them. Hot air will take days to penetrate a pile of clothes on the floor, but minutes to get between clothes spaced separately on hangers. Mattresses must have spaces between them and large fabric items like sheets or towels should be taken to the laundry in sealed bin liners and either hot washed or tumble dried above 60 °C. Curtains should be partly drawn so as to eliminate insulating folds, but not fully drawn so that insects can survive on the cool glass of the windows. Windows should be checked after the job and any survivors killed with insecticidal aerosol. Temperature sensors should be placed between or inside mattresses and furniture cushions, under rugs and carpets, inside any cracks in walls and behind wardrobes.

Damage may be caused to wooden laminates and antique furniture and to anything containing lacquer or glue that may melt. Paintings and paper are vulnerable, but most electrical items are surprisingly hardy provided that they are switched off. If in any doubt, it is best to carry out a visual inspection, treat the item with insecticidal aerosol and remove it, or ask the client to accept the risk.

In all cases, evaporation of water sucks out huge amounts of energy from a treatment, and in buildings the most obvious source of exposed water is in toilets. These should be covered with cling film to prevent evaporation of water, both from the point of view of energy conservation and to prevent over-humidification of the air. Where the air is too dry, some degree of water evaporation may be permitted, but the relative humidity (rH) of the air should ideally be monitored periodically during the course of the treatment to help prevent either excess drying or excess humidification, either of which can cause damage. It is of course important to remember to remove cling film from toilet seats after treatment.

Fleas are a common problem in residential houses. As with bedbugs, the type of flea should be ascertained first, not forgetting that bird fleas from attic-nesting feral pigeons, swifts, starlings, sparrows or jackdaws can be an issue, along with the full range of nest-dwelling insect pests. However, as fleas only occur in nest-dwelling animals that return time and time again to the same place (unlike deer and cows for example which bed down for the night in any place in woods or fields) for the treatment of the more usual dog and cat flea infestations the key to a successful treatment is to find and treat the 'nesting area'. This may be a dog basket, the owner's bed or even a habitual place where a cat lies down in the sun on the carpet or the dog in front of the fire. Such nesting areas can be vacuumed and sprayed to remove insects and eggs, which may render heat treatment of the whole house unnecessary. Treating the animal with prescribed flea medicine can also help eliminate such infestations.

Heat treatment against fleas can be very effective and may be the only way to kill eggs and pupae. As the infestations are inevitably at floor level, sheets maybe placed over the room furniture and carpets, and heat applied under the sheets. This concentrates the energy where it is needed and prevents it rising quickly to high wall and ceiling areas thus reducing risk of damage to these vulnerable areas.

Bed mites occur in beds and are usually just another name for house dust mites, however for over 30 years now it has been known that they are largely responsible for many asthma problems. All mites need warm, humid conditions to survive, along with a food source. Some, like bird mites, feed on blood, but house dust mites love the protein provided by human skin. Large amounts of skin rub off in our beds, and with humid body warmth and particularly on pillows where warm air is exhaled, conditions are perfect for mites to breed in large numbers. Heat treatment of bedding and mattresses has the double effect of killing mites and their eggs and instar stages, but also dehumidifying and drying the mattress to reduce risk of reinfestation for a while. It may be possible to argue that heat treatment also denatures the proteins in the dust mite faeces that is the cause of the allergic reaction that is asthma. Regular heat treatment of bedding could perhaps alleviate asthma allergies, but more research is needed in this area.

8.6 Summary

In summary, heat treatment of small rooms may seem simple, but it should be viewed in conjunction with the wider picture and as with all heat treatments, ideally used in conjunction with the full range of supplementary treatments, both chemical and physical.

References and further reading

Bell, C. (2014), 'Pest control of stored food products: insects and mites', in Lelieveld, H., Holah, J. and Napper, D. (Eds.), *Hygiene in food processing: principles and practice* (Second edition), Woodhead Publishing Limited, Cambridge, UK.

Heaps, J. (Ed.) (2006), *Insect management for food storage and processing* (Second edition), AACC Publishers, USA

Hill, D. (2002), *Pests of stored foodstuffs and their control*, Springer Publishers, The Netherlands.

Kloosterman, L. and Mager, K. (2014), 'Pest control in food businesses: an introduction', in Lelieveld, H., Holah, J. and Napper, D. (Eds.), *Hygiene in food processing: principles and practice* (Second edition), Woodhead Publishing Limited, Cambridge, UK.

Morland, D. and Hedoes, S. (Eds.) (2004), *The Mallis handbook of pest control* (Ninth edition), GIE Media Inc., USA

Subramanvam, B. (Ed.) (1995), *Integrated management of insects in stored products*, CRC Press, Boca Raton, USA.

Heat treatment of insect pests in transportation

9

9.1 Introduction

Heat treatment in transportation is not currently used as much as it could be, and yet it has the potential to replace most other forms of insect control treatment, including some commodity treatments. The reason for his is that transportation takes time, which is the biggest obstacle to heat treatments of commodities and difficult-to-penetrate cargo. Whilst current standard transport containers are not suitable for heat treatment, it would be simple to make adapted containers that allow for the free circulation of hot air. Just as we have refrigerated containers (reefers) that plug into an electricity supply for the duration of their jouney, so we could have heated containers working similarly for heat treatment. This chapter addresses the use of heat treatment to deal with insect pests in transportation. It begins with the use of heat treatment to deal with insect pests in transportation. It begns with the use of heat treatment in road transportation and trains, and the then discusses heat treatment of aircraft.

9.2 Heat treatment of road transportation and trains: overview

Heat treatment of insect pests in vehicles is becoming a significant issue, for example in preventing the movement of species between countries. As an example, if it were to enter New Zealand or Australia via the trade in automobiles, the gypsy moth would cause considerable damage to the forests in these countries if it were to become established. As a result, all vehicles entering these countries are inspected prior to import and either fumigated or heat treated if they are thought to either be infested or at serious risk of infestation. Another source of potential problems in vehicles has been that of NATO forces returning from Iraq and Afghanistan with vehicles and personal possessions infested with insects such as red-backed spiders (the black widow spider). Cars may also become infested with fleas or cockroaches as well as more exotic insects. Lorries with sleeper cabs and caravans of various kinds are also at risk of bedbug infestation.

Heat treatment of vehicles, whether cars, lorries, public service vehicles (PSVs) or railway carriages, is relatively straightforward and is one of the few types of heat treatment where a full survey may not be needed. The technician assessing the job needs to ascertain the type of vehicle, its location and position with reference to nearby power supplies. It goes without saying that gas burners should not be used near heated cars due to risk of ignition of fuel fumes. The type of pest should be ascertained, as recovering bodies provides evidence that the treatment has worked.

Heat Treatment for Insect Control. http://dx.doi.org/10.1533/9780857097811.3.83

9.3 Heat treatment of road transportation and trains: process

The sequence of events may be as follows:

1. Enquiry taken, establishing size and details of vehicle and its location. The client must be instructed to try to minimise fuel amounts onboard, particularly in petrol vehicles (diesel is less of an issue) and to remove fire extinguishers and all aerosols including de-icer aerosols and pressurised tyre inflation kits. The location of the nearest electricity supply should be established, or the heat treatment technician should take his own generator.
2. Agree the time and location of the job and make allowances for the weather. As the vehicle will need to be sheeted over, adverse conditions such as rain, snow and wind may affect the job. Small cars, trucks and buses may be double-sheeted during cold periods, and as the heat treatment van or trailer may be parked alongside the vehicle, some degree of recirculation may be possible in non-petrol powered vehicles. Full recirculation or non-convection heat treatment is NOT recommended due to risk of build up of fuel fumes in conjunction with a non-IP65 possible source of ignition.
3. Produce safe working practice and risk assessment, with risks being damage to any glues (particularly in caravans and motor homes), pressure vessels, owners' possessions such as cassette tapes or tins of food, and evaporating fuel.
4. There are several different techniques that can be employed for sheeting over cars, lorries and PSVs. They may be sheeted over individually, driven into pre-formed structures or poly tunnels, or treated *en masse* in heated buildings or garages. However, there are some caveats. First, the heat treatment should use either one-way heat exchange units (never direct burn propane) or *in situ* electric heaters or radiator systems, due to the risk of fuel fumes being evaporated during the heating process and building up concentration in a recirculation situation. The vehicle should always be sheeted over and insulated to avoid the outer metal skin taking on the average of the internal and external temperatures, similar to silos which must also be sheeted. As with silos, the only exceptions are pre-insulated vehicles such as refrigerated container lorries.
5. For vehicles, temperature sensors should be placed in the coldest places on the engine block (which is usually the last part of the metalwork to heat up due to its large mass), and inside seats and cushions and in floor wells. Blankets, bedding and clothing or personal effects should be removed and hot washed.

An example of the heat treatment of an automobile is shown in Figure 9.1. There are few potential hazards associated with heat treatment of vehicles other than pressure vessels and the risk of recirculation of petrol vapour. Even high-end entertainment systems seem unaffected by exposure to the 50–60 °C range of temperatures, but the usual rule of power off before treatment starts is as important here as in other heat treatments.

One other potential use for heat treatment may be in the deodorisation of cars. As an example, a three hour heat treatment with forced air ventilation at 60 °C evaporated and has dispersed substances causing a series odour problem. Similar work has been carried out in Germany to public service buildings to help reduce pesticide residues following timber treatments.

9.4 Heat treatment of aircraft: overview

Commercial aircraft have their own range of pest problems. Almost since long-haul air travel began there have been worries about malaria-bearing mosquitos spreading from Asia, Africa and South America to temperate regions. Airlines and government agencies now require application of insecticide (previously DDT, now replaced by modern pyrethoids) to eliminate unwanted stowaways. However, insecticidal sprays are not always effective when it comes to dealing with deeply embedded insect problems.

There are five main groups of potential insect/arthropod which may pose an infestation risk:

1. Bedbugs, fleas, ticks and other crawling biting insects.
2. Mosquitos, flies and midges and other flying biting insects.
3. Fruit flies and booklice.
4. Cockroaches
5. 'Casual intruder' insects, spiders and scorpions.

In order to control or prevent infestations we need to understand how insects develop and thrive. Bedbugs, fleas, ticks and other biting crawling insects may be carried by passengers on their clothing or in hand luggage and may bite other passengers, but they are unlikely to thrive, survive for long or maintain any form of resident viable population. Bedbugs, on the other hand, are beginning to represent a significant prob-

Figure 9.1 Heat treatment of insect pests in an automobile.

lem, as they are able to very successfully develop and sustain aggressive populations in even the cleanest aircraft of the most high profile western airlines, with hundreds of bugs present in some badly infested seats. Bedbug infestations are most likely to be found in the seams of cushions and under the Velcro-secured covers of the seats. Crew quarters in wide bodied jets may also become infested and become an issue.

There are three possible technical solutions to the problem of resident bed bugs in aircraft:

1. Fumigation – with either methyl bromide or CO_2 or potentially sulfuryl fluoride.
2. Heat treatment to a minimum temperature of 52 °C for two hours in the coldest place.
3. Residual insecticidal spray and dust treatment of potential bedbug harbourages.

Fumigation with methyl bromide is unacceptable in many countries on environmental grounds as it is a serious ozone depleter, and even then fumigations take 24 hours. CO_2 fumigations take even longer and sulfuryl fluoride has questionable efficacy against insect eggs at normal fumigation dosages.

Residual insecticides are suitable for use provided that all seats are treated under the covers where passengers' skin cannot come into contact with them. Diatomaceous earth applications may also be of assistance here. However, there is only a limited range of insecticides permitted for use on aircraft. This leaves heat treatment as the normal first choice option for aircraft infestations.

Technically heat treatment of aircraft should be simple (see Figure 9.2). An aircraft is essentially a well insulated metal cylinder full of lightweight metal based equipment covered in comparatively thin cushioning and fabric. There are restrictions on the maximum permitted temperature inside the aircraft. While different aircraft have different maximum operating temperatures, it is strongly recommended that the maximum air temperature is kept below 60 °C. Mesh covers in overhead compartments can melt at quite low temperatures, and temperatures of 70 °C may affect the operation or lifespan of some entertainment systems or other electronic components. As always, heat treatments should be carried out carefully and with due consideration for the potential for heat damage, and it is worth bearing in mind that damage to aircraft could have more serious consequences than a production line failing.

9.5 Heat treatment of aircraft: process

There are several possible ways of applying heat, but in the author's view the ideal solution should be as follows:

- Residual insecticidal treatment to all non-passenger contact surfaces with a permitted residual water-based insecticide and/or diatomaceous earth treatment. Exposed surfaces of seat covers, cushions and head rests must never be treated. Some people are highly susceptible to even the slightest insecticidal contact, and a serious medical emergency could result in mid-air if an allergic reaction were to occur. Often on hot long-haul flights babies may be undressed by their parents and come into contact with seat covers.
- All blankets and booklets should be removed along with all life jackets and fire extinguishers. Demo kits and any other potential harbour areas should be removed or treated separately. Blankets should be bagged and hot washed above 60 °C before being returned to the plane. Unfortunately, bedbugs may survive in the folds of life jackets and a decision needs to be taken on how to inspect/treat these without risking heat setting off compressed air cylinders. One solu-

tion would be to take them away and freeze or fumigate them with CO_2 in a chamber, replacing potentially infested jackets with sterile ones from the pool of chamber-treated life jackets.

- As heat has a tendency to rise, the seats should be sheeted over and heat applied below the seats. Powerful fans or ducted heat should be used to ensure an even, forced ventilation, which will make its own way above the sheets for an even distribution of heat. Whether glycol radiator systems and fans are used, hot air ducted from an outside source or electric heaters placed in the aircraft makes little difference – the only important technical considerations are heat distribution, observance of maximum and minimum air temperatures and time for heat to penetrate. It is often a good plan to heat the cargo hold as well, so that the floor heats up. Insects can easily survive on cold floors which are cooled from below by cold air. If the cargo hold cannot be heated, the metal temperature of its ceiling (the floor of the cabin) should be measured to ensure that the floor above has reached lethal temperature. Direct burn propane gas burners should never be used for heat treatment.
- Temperature measurement is essential. Sensors should be placed inside cushions and at floor level in the corners and middle. Core sensors should ideally be placed every 10 m across an aircraft, with floor and seat cushion locations down each side and in the centre (four abreast in wide-bodied jets; three abreast in smaller aircraft) with sensors in overhead lockers every 10 to 15 m. This will provide a good indication of genuine air temperatures. As a 30 m long section of a Boeing 747 cabin requires 12 low sensors (start and end 4 across at 15 m intervals) and 12 high level sensors (again 4 across at 15 m intervals), plus air in and air out sensors, making some 26+ sensors in total, it is recommended to use radio linked temperature monitoring systems to minimise cable issues, making sure of course that all aircraft systems are shut down and that the system uses a frequency and power rating approved for use in an airside environment. Infra-red spot meters may be used to measure the hold ceiling temperatures during heat treatment if the hold is not separately treated, but of course cannot measure trends and rates of temperature increase like a data logging system can.
- A target of 52 °C for two hours is more than sufficient to kill all stages of the bedbug life cycle, provided measurements are taken on the insect rather than spot metering metal or surface temperatures. As with all heat treatments, it is the real core temperature that counts, not surface temperatures, as these can be directly influenced by hot air.
- Aircraft systems are tested and certified to be able to stand temperatures of 65–70 °C as they have to be able to operate in the hottest desert conditions with a margin to spare. Cars and vehicles stand for hours in the hottest direct sunlight without ill effect so are quite resistant to heat treatment, and the conduction of heat along metal makes for good heat penetration and distribution.
- Crew accommodation should not be ignored and is often a source of persistent bedbug infestation.

Mosquitoes, flies, midges and other flying insects are easily killed by heat, requiring lower temperatures of just 50 °C for one hour. It is not necessary to sheet over the seats, but ducted air should be kept low to kill fallen insects. As these insects do not breed in the aircraft, there is no worry about eggs or pupae.

Fruit flies and booklice tend to infest areas prone to sticky food and juice spillages, such as galley areas. It is perfectly acceptable to treat just these areas to kill insects and to help dry out spilled juices and beer in inaccessible areas. However, as always with these types of infestation, detailed cleaning is the long term answer to remove food sources and breeding harbourages.

Cockroaches on the other hand can be a serious pest as they are highly visible and can represent a food hygiene hazard, particularly relating to *Salmonellae* infection. However, unlike many other insects, cockroaches tend not to be brought on by pas-

Figure 9.2 Heat treatment of insect pests in an aircraft.

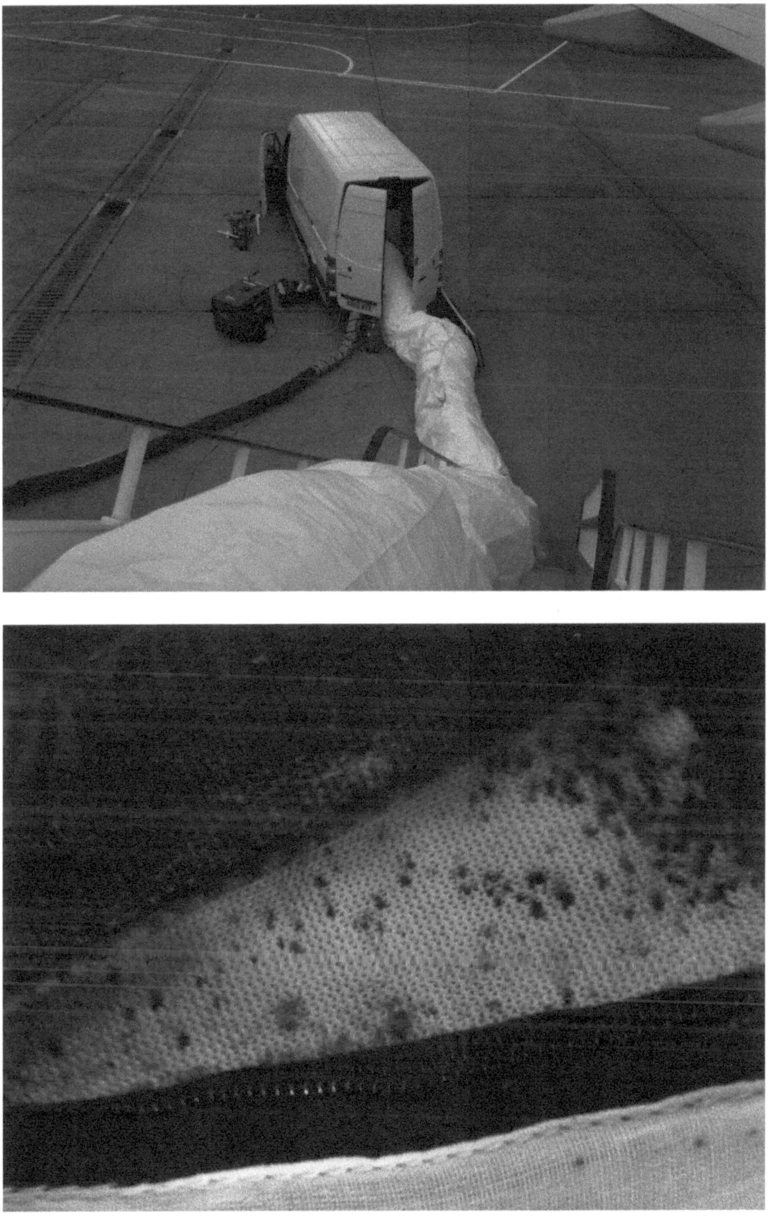

Figure 9.2 Continued.

sengers but by caterers. Some airline food production and dishwashing facilities can become heavily infested with cockroaches, as proper treatment of infested airline catering and industrial dishwashing facilities requires downtime, which may not be built into the system. Regional pest controllers do their best to control cockroaches in these facilities with insecticidal gels, baits and residual sprays of all sorts, as well as the hot water and steam washes necessary for food hygiene, but cockroaches can be a persistent and hard to eradicate hazard.

Fortunately cockroaches and other casual intruder insects are also susceptible to heat treatment, requiring the same temperature × time product as flying insects. Cockroaches and other crawling insects respond extremely well to combined heat and residual insecticidal treatments, where the heat forces them over insecticide-treated surfaces that they might normally sense and avoid. German cockroaches (*Blattella germanica*) can also be killed using low temperatures, with temperatures of 2 °C or less for eight hours often proving fatal. Oriental cockroaches (*Blatta orientalis*) have been recorded walking over snow and can seem pretty indestructible to most forms of pest control measures at times – but as with all insects, heat treatment will kill them eventually as long as they cannot find refuge.

9.6 Summary

Being largely metal, vehicles are highly suited to heat treatment. In addition, many are highly unsuitable for phosphine fumigation, particularly as phytosanitary treatments often take place in port areas, which have humid atmospheres that enhance the tarnishing of conductive metals by phosphine gas in electronic components. The only drawback is the sheer volume of vehicles to be treated, requiring large treatment facilities. However, treatments can, for example, be carried out in transit in the holds of car transporter ships, without the risks associated with in-transit phosphine fumigation. All that is required for in-transit jobs is bigger heaters and the power to run them. As with all forms of pest control, long-term insect control and prevention in aircraft requires thought, planning and good integration of different processes and technologies.

References and further reading

Bell, C. (2014), 'Pest control of stored food products: insects and mites', in Lelieveld, H., Holah, J. and Napper, D. (Eds.), *Hygiene in food processing: principles and practice* (Second edition), Woodhead Publishing Limited, Cambridge, UK.

Heaps, J. (Ed.) (2006), *Insect management for food storage and processing* (Second edition), AACC Publishers, USA

Hill, D. (2002), *Pests of stored foodstuffs and their control*, Springer Publishers, The Netherlands.

Kloosterman, L. and Mager, K. (2014), 'Pest control in food businesses: an introduction', in Lelieveld, H., Holah, J. and Napper, D. (Eds.), *Hygiene in food processing: principles and practice* (Second edition), Woodhead Publishing Limited, Cambridge, UK.

Morland, D. and Hedoes, S. (Eds.) (2004), *The Mallis handbook of pest control* (Ninth edition), GIE Media Inc., USA

Subramanvam, B. (Ed.) (1995), *Integrated management of insects in stored products*, CRC Press, Boca Raton, USA.

Heat treatment of insect pests in clothing

10

10.1 Introduction

Although this problem has decreased over the years, pest controllers still come across plenty of cases of damage to clothing by the usual textile pest suspects: carpet beetles (*Anthrenus* spp), hide beetles (*Dermestes* spp), fur beetles (*Atttagenus* spp), clothes moths (*Tinea* spp and *Tineola* spp), as well as white shouldered house moths (*Endrosis sarcitrella*). The initial source may range from a bird's nest to an article of clothing bought or borrowed from an infested source.

Single infested items of clothing may be hot washed or frozen, but pest controllers may be brought in when infestations in large collections of clothes have gone unnoticed and got out of hand, such as museum costume collections, clothing retailers, costume hire or simply members of the public with a large number of clothes. Typical types of clothing that become infested are woollen items, fur coats and animal skins, or even stuffed animals. Treatment of clothes is simple, but requires more time than heat treatment of machinery. The clothing needs to be separated out and hung up so that hot air can circulate freely around each individual item and penetrate to kill all insect life stages.

10.2 Heat treatments

As some items of clothing can be delicate, it is essential to always check which heat parameters are acceptable to the client. However, heat alone will not damage many items of clothing. For example, it is not recommended to hot wash woollens, this is because washing is often accompanied by harsh stretching actions in the washing machine, which can cause stretching of weakened protein fibres. With heat treatment there is no motion, and to date the author has not experienced any problems with heat treating woollens to 52 °C. Longer exposures at lower temperatures of 48 °C allow for less extreme temperatures and of course more time for better penetration of heat into difficult areas. Bales of clothing cannot reasonably be heat treated except by long-term exposure in heat treatment chambers, where heat treatment times may be comparable to that of fumigation. In such cases, fumigation with phosphine may be preferable and cheaper.

One solution is a polytunnel made specifically for treating potentially infested clothing. Hot air is blown along the base of the tunnel beneath the suspended and

Heat Treatment for Insect Control. http://dx.doi.org/10.1533/9780857097811.3.91

separated clothes and perforations in the top of the tunnel allow heat to pass upwards through the clothes. Frequent temperature monitoring with sensors inside pockets and linings will allow the heat treatment technician to heat all items up to 52 °C for a minimum of two hours, without exceeding 60 °C anywhere.

10.3 Key steps in heat treatment

A typical heat treatment operation will have the following components:

1. Initial survey, where the surveyor will itemise the number and nature of the clothing that needs to be treated, bearing in mind that he will be aiming to treat the juvenile stages of the infestation (eggs, larvae and pupae) with heat. Visible adults in the fabric of the building can be sprayed, fogged or even treated with aerosols if label approval permits. Clothes made of synthetic fibres are unlikely to be harbouring eggs and larvae, although the presence of pupal stages is not impossible, depending on the location of the items and their proximity to infested natural fibres. The surveyor should also determine how the infestation arrived on the premises. Did it come in with infested stock or returned items, or is it endemic on the site, e.g. having originated in a bird's nest. Attic areas and the outside of the building should always be checked for nests.
2. Planning survey, where the operator in charge of the job or a trusted representative will work out how the job is to be carried out, whether it is possible to treat wardrobe areas *in situ* or whether a heat treatment polytunnel needs to be constructed. In some cases, a temporary heat treatment structure may be constructed incorporating a wall of the house or retail outlet.
3. Using the results of the above survey, a safe working practice or operating procedure will be developed and agreed with the customer. Here the customer will be informed of any potential risks to his product and have a chance to ask questions. Whilst glues are not normal in items of clothing, operators should be careful of shoes that may have a glue element, and if incorporating parts of walls or wardrobes be aware of any ornamental components glued on to doors, etc.
4. Treatments will often include an additional insecticidal component to treat any wandering adult moths, in order to prevent re-infestation of the treatment area. If the client is concerned about toxic insecticides, fogging with natural pyrethroids may be acceptable.
5. As with bedbugs in bedding, penetration of heat between insulated layers of clothing is a big issue, and the spacing and hanging of the clothes to be treated is important. It is also important that air circulation between items is maintained. The technician should enter the stack and move the clothes around every hour or so to ensure that no cold spots are allowed to develop.
6. Prevention of damage is crucial, and lower air input temperatures should be used for longer exposure times, aiming for core temperatures of 52 °C for two hours without exceeding 60 °C.
7. In high value jobs, it is always wise to keep the client involved and give them the chance to inspect items as the treatment progresses. This gives the client the option of calling a halt to the treatment at any time should they be concerned.

Pheromone monitoring systems are available for many types of textile pests, and these may be used both to help pinpoint the core of the infestation and in monitoring the success or otherwise of the treatment.

One of the bigger problems facing pest controllers trying to control textile pests or indeed bedbugs in domestic situations can be the sheer mass of personal clothing and personal effects. To complete a heat treatment job properly under these circumstances, clothing may be hot washed or tumble dried by the client or frozen for one to two weeks in a deep freeze. With freezing, remember that just as heat takes a long time to penetrate a thick bale of clothing, so does cold.

As with all pest control jobs, a correctly performed survey is crucial and the proposed solution should be holistic and all-encompassing. Heat treatment should be supported by other insect control measures as appropriate and the client must be educated and encouraged to do his part as well.

References and further reading

Bell, C. (2014), 'Pest control of stored food products: insects and mites', in Lelieveld, H., Holah, J. and Napper, D. (Eds.), *Hygiene in food processing: principles and practice* (Second edition), Woodhead Publishing Limited, Cambridge, UK.

Heaps, J. (Ed.) (2006), *Insect management for food storage and processing* (Second edition), AACC Publishers, USA.

Hill, D. (2002), *Pests of stored foodstuffs and their control*, Springer Publishers, The Netherlands.

Kloosterman, L. and Mager, K. (2014), 'Pest control in food businesses: an introduction', in Lelieveld, H., Holah, J. and Napper, D. (Eds.), *Hygiene in food processing: principles and practice* (Second edition), Woodhead Publishing Limited, Cambridge, UK.

Morland, D. and Hedges, S. (Eds.) (2004), *The Mallis handbook of pest control* (Ninth edition), GIE Media Inc., USA

Subramanvam, B. (Ed.) (1995), *Integrated management of insects in stored products*, CRC Press, Boca Raton, USA.

Index

I

Kolibri 360 GmbH
Ferdinand-Jühlke-Straße 7
99095 Erfurt, Deutschland
produktsicherheit@kolibri360.de